Your QRP Operating Companion

By Brad Wells, KR7L

Published by
The American Radio Relay League
225 Main Street, Newington, CT 06111

Copyright © 1992 by

The American Radio Relay League Inc

Copyright secured under the
Pan-American Convention

This work is publication No.150 of the Radio Amateur's Library, published by the League. All rights reserved. No part of this work may be reproduced in any form except by written permission of the publisher. All rights of translation are reserved.

Printed in USA

Quedan reservados todos los derechos

$6.00 in USA

ISBN: 0-87259-376-2

First Edition

Cover photo by Kirk Kleinschmidt, NTØZ

Table of Contents

Acknowledgments

Preface

Chapter 1   Why Run Less Than The Legal Limit?

Chapter 2   What Do I Need?

Chapter 3   Operating Techniques

Chapter 4   Maximize Your Signal

Chapter 5   A Propagation Primer

Fadeout

Appendix A   QRP Clubs and Nets

Appendix B   Contests With QRP Categories

ACKNOWLEDGMENTS

No author works in a vacuum when writing a manuscript. Any book is laced with the thoughts, ideas, convictions, and advice of other people. Some of those who contributed to this book are:

My wife Cathy, N7KOV, whose comments and criticisms during the editing process have been incorporated into the manuscript;

Mark Wilson, AA2Z, Managing Editor of *QST*, who gave me an insider's look into the publishing business at the 1990 Seaside Hamfair;

Jim Kearman, KR1S, Assistant Technical Editor, who helped smooth out many of the rough spots in the original manuscript and provided the insight of an active QRPer;

Jack Bock, K7ZR, Ed Aho, NR7F, and other members of the Western Washington DX Club who listen to my 5-watt exploits and spot the rare ones, but don't hesitate to bury me in a pile-up.

PREFACE

The capture and confirmation of rare countries is a ham radio passion contracted at any stage of the sunspot cycle. I first became interested in this mode of operation in the waning years of Solar Cycle 21. Everybody knew that DX could only be worked with an amplifier during those barren years of low solar flux.

Following the conventional wisdom of the day, I operated with a kilowatt amplifier. However, I began to get a little bit tired of the neighbors' complaints while I worked my fair share of DX. We darn near had a divorce in the family when it was discovered that DXing, 1000 watts, and cable TV are not really compatible. Something had to be done.

So we moved to the country. Unfortunately, all these problems just seemed to follow right along. The solar flux didn't improve and the televisions didn't disappear. The domestic problems were further compounded by my growing interest in contesting.

At first, I shunned this weekend warfare. After all, the 48 hours of screaming chaos on every band was an assault on my finer sensibilities. Then, repugnance gave way to curiosity. I discovered that rare DX stations, not normally heard, operated on these Saturdays and Sundays. Even more interesting was the fact that they were easy to snag and stick in my log book. I got to looking at contest scores in the major magazines. Do you know that some of these operators were racking up over 3000 contacts in a single weekend? I paid more attention to the photographs of their stations and discovered something else.

Radio contesting is strictly a root-hog-or-die proposition. Winning stations have racks of amplifiers and glamorous operating positions. This is really kilowatt country. It's 1500 watts or nothing. Well, almost nothing. I found that some of these contests actually let you operate in a category where you had to run 5 watts or less output. The amazing thing is that they were DX contests. Who ever heard of working DX with 5 watts?

Now, I'd worked an occasional QRPer while chasing DX up and down the bands. They always struck me as poverty operators forced to be content with miniscule rigs since they couldn't afford real radios. They always seemed so incredibly happy to make any sort of contact. But low power looked like a possible solution for the growing domestic and neighborhood tensions. I turned down the drive control on my rig, unplugged the amplifier, and jumped into the CQ World-Wide Contest.

I figured my best DX might be a KL7 or, with luck, a KH6. Now, I'd like to brag that my first QRP contest entry was something really spectacular. It wasn't. But I did work some DX including a KL7 and KH6, and guess what? The family and neighbors had absolutely no idea that I was on-the-air for almost 24 hours during that weekend. I was hooked.

Like anyone else interested in a new facet of Amateur Radio, I went looking for some more information. What did I find? Nobody had written any books and there were nothing but construction articles in all the magazines. It seemed like everybody already knew how to operate QRP, they just didn't know how to build the rigs. Since nobody was talking about how to use five watts, I launched out on a program of self-education. You can't help but learn something if you radiate a little RF energy in every direction.

For example, I discovered that when operating with 5 watts or 1 watt or 100 milliwatts, the conventional wisdom doesn't always work. Chasing and catching a KH1, a ZD7, or TA2 in a pile-up when you're running 5 watts is just a bit different than when you're stoking 1500 watts up the coax. Likewise, the exchange ends up a bit different if you're calling a 1-watt CQ during a major DX contest. Everyone answers with: "Is This Frequency In Use?"

This book is a distillation of my QRP operating experiences. These past years of running low power have given me some very definite opinions and personal bias on what works and what fails. I've developed a real empathy for the Suffering Sevens. If there's a black hole in the ionosphere, it's located directly over the Pacific Northwest. Seems like we're always the last to get propagation to anywhere.

If you try QRP, and stick with it, you can't help but become a better operator. It's one of the few endeavors within Amateur Radio where expertise has little or nothing to do with available hardware. Let's set the record straight: It's a lot easier to be a "good" operator if you're pumping 1500 watts to a 6-element monobander mounted on a 100-foot tower than if you're running 1 watt to a ground-mounted vertical. In the long run, the QRPer can work the same rare DX as his QRO counterpart. However, along the way, the QRP operator will have developed a deeper reservoir of operating skills and propagation knowledge. Most certainly, they will have learned the values of patience and perseverance.

I hope this book can reduce the learning curve for most of you. I've written it on the assumption that you already have some DXing experience running barefoot and, perhaps, have dabbled in a contest or two. Hopefully, you won't have to spend the next several years learning to be an expert low power operator. QRP becomes more enjoyable, and a lot less frustrating, after you learn some of its specialized operating techniques.

QRPers tend to fall into several diverse groups. There are the construction buffs who relish the challenge of building an all-mode, 1-watt rig the size of a cocktail sugar cube. There is another group to whom QRP represents a laid-back mode of operation and they relish equally each and every contact, whether it be local or DX. Then, there's the group to which I belong: QRPers who see themselves as the guerrilla operators of the ham bands. You will notice that parts of some chapters read like a combat manual. There is a reason for this. Every pile-up is a war zone littered with casualties and filled with the sounds of agony. Victory goes, not to the strongest, but to the smartest. You must squeeze the maximum effectiveness out of your few watts in this battlefield of the ionosphere. In short, you must be doing it cleaner, faster, and better, much better than the other operator. DXCC and contests provide no plaques or trophies for the second-place winners.

The one thing this book can't teach you is patience or perseverance. You're going to have to develop those skills all by yourself. In the meantime, get out there and radiate a few watts of RF. There's a brave new world of operating experiences waiting just beyond your antenna.

You can read the chapters of this book in any order that suits you. While they dovetail one into the other, they are also independent of each other. You will find similar themes popping up in the various chapters. They are the hallmarks of successful QRP operation. A fast reader could probably digest this book in one evening. I hope that you don't. The real feel for QRP will come only if you read at a slower rate and try to apply the techniques to your own style of operating.

I measure QRP success by the cards on the wall, the contest scores, and the certificates earned. They are often unrelated to the amount of dollars you've invested in equipment. All successful QRP operators do one thing in common, regardless of what rig, mode, or antenna is being used. They set attainable, personal goals. QRPers realize that they are not competing with each other, but only against themselves. QSL cards and wallpaper provide solid proof that they've pushed right to the edge of their operating envelope.

CHAPTER 1

Why Run Less Than The Legal Limit?

Deep within each of us is the desire to do or be different from everyone else. Running QRP (less than 5-watts output) is not for everyone. A bit of luck and a lot of skill are the determining factors in your ultimate success. Working the world with 5 watts demonstrates both your abilities and operating skills, and it's a lot of fun besides!

There are many reasons why you might choose to run only 5 watts. The obvious reason is the challenge of succeeding in a world where most hams run at least 20 times more power. Another reason may be the interference problems faced by hams who live in cities or suburbs. If your phone rings every time you try to operate, it's time to consider QRP. Interference is seldom a problem for the QRPer. You can run your 5-watt rig day and night and the neighbors will never know.

Finally, you may have been using higher power and found it's too easy. Collecting QSL cards is lots of fun, but there isn't much challenge in always getting the DX on the first call!

Even if you're an experienced and successful DXer or contester, if you've been running 100 watts or more until now, you'll find good information in this book. The louder you are, the less need you have for strategy and tactics. In fact, applying the ideas contained in this book when you're running high power will bring amazing results. So don't throw away the linear; just put it on standby.

Oh, by the way. You may vaguely remember the FCC rule regarding "the minimum power necessary to complete the commmunication." When you bust a pile up with 5 watts or less, surely everyone else calling will also reduce power, right? Well, probably not, but you can bet you caused a few more gray hairs to appear on certain DXers' heads!

One word of caution before you go on: QRP may take control of your Amateur Radio life. The Surgeon General should post warning labels on low power rigs:

WARNING

QRP MAY BE ADDICTIVE AND LEAD TO LOSS OF AMPLIFIERS

For whatever reason you begin to operate QRP, it is habit-forming and will forever change your operating style.

ISN'T QRP A BIG HANDICAP?

Five watts of RF energy doesn't sound like much to those who consider running 100 watts as low-power operating. Some hams positively insist that communications are impossible without an amplifier. In fact, 5 watts will little affect your ability to work DX. The difference between QRP and 100 or 1500 watts is only 3 to 5 S units. "How can I talk to anybody if I'm only S5?" you ask? There is absolutely nothing wrong with receiving a 539 signal report. With very little effort, you can copy stations that are so weak they don't even wiggle your S meter. We don't *need* telephone-quality communications, especially when running the necessary power to obtain them causes so much interference

QRP TRADITION

Most pioneer hams never put kilo before watts when describing their rigs. Most often, they were running a lot less than maximum legal power. Given the poor efficiency of many early rigs, the plate dissipation of their final amplifier tube was often greater than their output power. Adrian Weiss, WØRSP, has put together a fascinating history of early QRP operation, which you'll find very inspirational[1]. Unfortunately, what happened in Amateur Radio was the same thing that happened to the rest of our lives: Bigger is Better; Kilowatts beat Milliwatts; and everybody had to have one of those glistening, powerful rigs featured on the covers of amateur magazines. It suddenly became obvious to everyone that an S9+40 dB signal report was much better than an S9 report. We could no longer communicate without pinning somebody's S-meter.

The use of kilowatts for cross-town QSOs eventually triggered a counter-revolution. Many hams once again discovered that 55 reports don't mean an end to communication. This revolution has progressed to where it is now fashionable to operate with minimum power levels.

HISTORY OF QRP COMPETITION

Competition has been a part of Amateur Radio since the beginnings of the hobby. When the ARRL was founded in 1914, an amateur station with a range of 400 miles was doing well. In fact, the limited range of stations at the time was the incentive for forming the American Radio *Relay* League. Using relay stations enabled amateurs to bridge the North American continent. Amateur communications in those days took place on frequencies below 1500 kHz, in today's AM broadcast band. We now know that using higher frequencies would have given better results, but amateurs were reluctant to move to shorter wavelengths. To accomplish their distance goals, the successful stations simply ran as much power as possible. "Survival of the fittest. Higher and higher powers were the order of the day. The race was on, and devil take the hindmost. Interference. Lord, what interference! Bedlam![2] Sound familiar?

One-thousand-watts input to a spark station rarely netted

contacts beyond 1000 miles, although some spark stations were heard in Europe during the first Transatlantic Tests. By the time of the Tests, though, coherer and crystal detectors were replaced by regenerative vacuum-tube receivers. The application of vacuum tubes to transmitting really increased the distance a station could cover. Vacuum tube oscillators, by concentrating their power on a single frequency (a "tuned" spark transmitter had a bandwidth of about 20 kHz!) were far more efficient than even the best spark rigs.

The problem with using continuous-wave (CW) transmitters was the lack of frequency stability. Neither the regenerative (oscillating) detectors used in receivers nor the free-running transmitter oscillators were especially stable. Receiver front panels were made from non-conductive materials. Moving your hand near the tuning and regeneration controls to compensate for a drifting transmitter often detuned the receiver enough to lose the signal altogether.

The first amateur CW transmitters used receiving tubes. The plate dissipation of these tubes was specified as about 5 watts, but amateurs often abused them unmercifully. A "5-watt tube" might produce as much as 100 watts. Before long, tubes designed for transmitting became available to amateurs, and the power race was on again.

The "Best" Stations

In 1921, spark stations still snarled and spat, but CW was rapidly forcing its way onto the bands. Some stations had both. Here's a description of the "best amateur station" of 1921:

"1. A 200-watt *(plate dissipation; who knows how much power it could run)* CW, ICW and radiotelephone transmitter, utilizing four 50-watt tubes and one 5-watt speech amplifier tube, built on a bakelite panel mounted on an aluminum frame, power by two 200-watt 500-volt motor-generators sets in series to give 1000 volts.

2. A 1-kilowatt rotary spark transmitter, utilizing duplicate transformers, a rotary gap driven by a synchronous motor, a

copper-strip oscillation transformer, an oil-immersed condensor, and a thermocouple radiation (antenna-current) meter.

3. A switchboard controlling all power circuits and enabling any part of the station to be switched in or out of operation.

4. A receiving cabinet, consisting of a variometer regenerative set with two stages of audio amplification.

5. An aerial system comprising a 4-wire 90-foot flat-top supported on masts 67 feet high, with 14-foot spreaders, and a fan-style downlead; a fan-type counterpoise consisting of six 50-foot wires; and a ground connection made to three buried hot-water tanks, to pipes, and to buried wires."[3]

The owner of this massive station, Louis Falconi, 5ZA, of Roswell, New Mexico, was obviously serious about getting out. This state-of-the-art station could be heard all over the continental United States and Hawaii. By noting the disparity between the high transmitter power and the regenerative receiver, we can guess that 5ZA could be heard by more stations than he could hear. The superheterodyne receiver had just appeared on the scene, but the regenerative receiver was not yet fully developed. Variometer (variable-inductance) tuning was difficult to adjust, but better receivers were yet to come. Spark didn't give up easily: The "best station" of 1922 was all-spark 2OM, in New Jersey.

Transoceanic Tests

If 5ZA could be heard on the East Coast, then transatlantic contacts by amateurs were a sure thing, or so it seems now. The amateurs of the time proceeded with what we would consider frustrating caution. ARRL determined to first conduct tests where American hams would transmit and European hams would listen. Before you could even participate, you had to prove your station by covering 1200 miles in preliminary tests! During these preliminary tests, 91 North American calls were logged in Europe. Still, the tests continued. When they were over, 316 American stations from all call areas had been heard in Europe. How could the Europeans resist calling back? Some did. About 20 American hams heard European amateurs. Despite these successes, the first amateur two-way contact between North America and Europe finally took place on November 27, 1923. The wavelength used was the relatively unexplored 100 meters (3 MHz).

Low-Power Tests Down Under

With the imaginary ocean wall bridged, contacts across the Atlantic and Pacific oceans became almost commonplace. The big dream of "getting over the pond" realized, some amateurs looked for other worlds to conquer.

"Perhaps the most amazing transmission feat of all time was performed by Charles D. MacLurcan, 2CM (prior to regular international amateur communication, callsign prefixes were unnecessary), of Sydney, Australia, on September 26, 1923. Following a series of tests with extremely low power, a schedule was arranged with Frank Bell, 4AA, of Waihemo, Shag Valley, New Zealand—1500 miles away. After communication had been established, the power input to the transmitter at 2CM was reduced to 0.7 watt. Signals remained good. The next step was 0.04 watt. Finally, satisfactory signals were received not only in New Zealand but 1400 miles north, in Queensland, when a power of *only 0.004 watt*—0.25 milliamperes at 15 volts—was used! This is a record [that] no one, amateur or professional, is likely ever to beat."[4]

DeSoto goes on to say that more than 100 Americans had been copied in New Zealand and Australia by the end of October 1923. Eventually, someone in North America heard and worked a station "down under." What was the problem? From the vantage point of time, it seems that US hams were obsessed with power. Station 6AWT reportedly ran 5400 watts input to a 250-watt tube. New Zealand's MacLurcan was not impressed. To see why North American hams couldn't copy him, he sailed to San Francisco, bringing his receiver along. From San Francisco Bay he succeeded in hearing his 8-watt transmitter back home. During these tests, no American stations were able to hear the same signal! MacLurcan acknowledged greater levels of interference in the US, but he put that down to the excessive power we ran as much as the number of stations active.

Station Efficiency

QST Technical Editor Kruse announced a "Station Efficiency" contest in the December 1923 issue.[5] The first results appeared in May 1924 *QST*.[6] The station-efficiency contest sought to maximize "miles per watt" input, rather than simply miles covered. Even 6AWT, using his other call of 6AWS, joined in. His best effort, with 16-watts input, got his signal to Hawaii, for 130 miles per watt. Canadian 3NI, on the other hand, managed 750 miles with 360-milliwatts (0.36-watt) *input*, for 2080 miles per watt. We can only wonder if 6AWS/6AWT got the message! By the way, the low-power reports had some words of advice that are worth repeating: "Four of the men . . . point out that in low-power work it is particularly necessary to call in an orderly fashion. No one will listen to a weak signal that makes 164 calls before signing. . . .Another thing useful in doing low-power work is picking out the proper time and wavelength for it. . . . The sole idea is to hit on a period when the air is least occupied and there is [the] best chance for a weak signal to make itself heard. Pick out a wave that is the freest from [Q]RM, no matter which of the wavebands you work in."

With no mandated power limits, the idea of QRP DX was born. Running low power in crowded bands was as much of a challenge then as it is now. *QST* published many articles on "low-loss" receivers and spectrally clean, stable transmitters. Again, like today, some hams got hooked on QRP, while others ran as much power as possible. A new attitude was developing, though. *QST* for June 1925 contained this note: "The Wouff-Hong trophy which is to be awarded to the best amateur station in the sixth district, Pacific Division ARRL, should have a whole flock of applicants. The award will be made annually starting with the November convention in Santa Barbara. There are four points which will be used in judging the stations: Log (35%), Consistency of operation (25%), DX in *miles per watt* (20%), and neatness and arrangement and percentage of 'home-made-ness' of apparatus (20%)."[7]

Contests

As distance tests evolved into contests, a separate category was established for "low-power stations" (usually 100-watts input), but QRP stations were ignored in the rules. Low-power stations were sometimes allowed to multiply their scores by some factor thought to make them equal to the high-power stations. This situation continued until fairly recently, when (by popular demand) separate classes were established for high-power, low-power and QRP stations.

The first major contest to separately recognize QRP operators in the results was the 1979 *CQ* World Wide DX Contest. Adrian Weiss, now WØRSP, sponsored plaques for top CW and SSB entrants. The following year, *QST* began publishing a separate QRP "Top Ten" in the ARRL DX Contest results. The ARRL Sweepstakes contest followed suit in 1987. Today, nearly every major contest recognizes QRP participation, and competition for top-ten scores is heavy.

What Is QRP In A Contest?

There is a subtle difference in the definition of QRP between contests sponsored by *CQ* and those sponsored by ARRL. For contest purposes, the ARRL defines QRP as 5-watts RF output, while *CQ* defines it as 5-watts DC input. Until recently, the British G-QRP Club allowed a maximum of 3-watts RF output to qualify for its awards and operating activities. G-QRP now uses the 5-watts RF output definition. The difference between 5-watts and 3-watts output is 2.2 dB, which *may* make a difference in some contacts.

Comparing Your Results With The "Big Guns"

While it may be possible to come up with multipliers that make all stations equal, it is much more sensible to let stations of equivalent power compete with each other. When you enter a contest in the QRP class now, you aren't competing with stations running 1500 or even 100 watts. You're competing with other QRP stations. After the contest, compare your QSOs per watt or points per watt with the high-power stations. You may never make as many contacts as the big guns, but you'll certainly be pleased with the relative performance of your QRP station.

WHEN NOT TO USE QRP

To be completely honest, there are several times and situations where QRP is out of place and not in your best interests. For example, handling message traffic requires a power margin that ensures error-free communications. There is nothing hit-or-miss about passing written messages. It isn't something that can be tried again tomorrow if the ionosphere fails to cooperate today. Likewise, emergency communications require enough power to guarantee complete reliability in difficult situations.

Personally, I don't recommend QRP to new hams. When you're just starting out, you need all the help you can get. This includes 100 watts of RF energy to the antenna. It's not that beginners can't work DX with 5 watts, but it's most important for them to have a low level of frustration. This healthy dose of wattage allows them to develop basic operating skills. Later on, low power hones them for you.

The basic definition of QRP, recognized by most amateur organizations, is 10-watts input or 5-watts output. However, the definition varies slightly between these groups. Some consider 10-watts PEP on SSB to be the equivalent of 5-watts out using CW. I consider QRP to be a maximum of 5-watts output, regardless of mode. I use a peak-reading wattmeter to measure power output no matter where I set the mode-selection switch.

Input power is seldom used these days to determine whether-or-not you're QRP. It's much easier, and far more accurate, to put a calibrated wattmeter in-line between the rig and the antenna.

QRPp: FOR THE TRUE BELIEVERS

If you really want a challenge, there is a sub-class of QRP known as QRPp. QRPp is defined as 1-watt or less output. Milliwatts place a premium on perseverance and operator skill[2]. The difference between running QRPp and QRP is very similar to the difference between running barefoot and with an amplifier.

Contacts and countries that were easy with 5 watts become a real challenge when you're only pumping milliwatts to the antenna. The first DXCC Milliwatt Award went to W8ILC in 1977, prior to the peak years of that solar cycle. Several years later, Ron pushed his country total above 300. For most contacts, he ran less than

1 watt on single sideband.

Don't expect other hams to be very impressed with your low-power exploits. The few who will appreciate your efforts are those other intrepid QRP types. Even your family will be little impressed with your 5-watt conquests. Running very low power is often an extreme exercise in humility. There's no afterburner to kick in and you won't stomp anybody in a pile-up. Most of the time, you'll end up as the stompee, not the stomper. There will be times when you drag a very bruised ego away from your radio. The ideas and techniques discussed in the following pages should reduce your annual expenditures for bandages, aspirin, and psychiatric treatment.

You'll be able to hold your head high while speaking proudly of each and every contact you've made. You can burn up the bands secure in the knowledge that your superior operating skill is more efficient, more rewarding, and more fun than using the maximum available hardware.

Notes

[1] A. Weiss, WØRSP, *History of QRP in the U.S., 1924-1960,* Vermillion, SD: Milliwatt Books, 1987.
[2] C. B. DeSoto, *200 Meters And Down*, p 60 (Newington: ARRL, 1936).
[3] *200 Meters And Down*, p 80.
[4] *200 Meters And Down*, p 85. Allowing for 65% final-amplifier efficiency, MacLurcan achieved 538,461.5 miles per watt. DeSoto underestimated QRPers! Gain antennas and favorable propagation have made 1-million-mile-per-watt contacts possible. MacLurcan's technique of making contact, then reducing power until the signal was lost, is not accepted by today's QRPers. To be valid, the *entire* contact must be made at the claimed power level.
[5] S. Kruse, "Miles Per Watt," *QST*, Dec 1923, pp 46-47.
[6] S. Kruse, "That 'Station Efficiency' Contest and the New American Amateur," *QST*, May 1924, pp 36-37. Follow-up reports appeared in *QST*: Dec 1924, pp 46 and Jun 1925, pp 44-45.
[7] Untitled, anonymous item, *QST*, Jun 1925, p 45.
[8] E. Hare, KA1CV, "Confessions of an Inveterate Milliwatter" *QST*, Feb 1990, p 44.

CHAPTER 2
QRP Equipment— What Do I Need?

There are dozens of QRP rigs available, new and used. Most are smaller than the average 100-watt-output transceiver, which makes them useful for backpacking or canoeing trips. However, to achieve a smaller package, something was obviously left out. With a few exceptions, QRP rigs don't have digital displays, speech processors or the variety of receiver I-F filters available on higher-powered transceivers. Especially in QRP contests, where you have to copy many weak signals, the average dedicated QRP transceiver is not up to the task.

One exception is Ten-Tec Argonaut II.[1] The Argonaut II has many of the features found in higher-powered transceivers, but its output power is at the QRP level. If you are sure you'll only want to run 5 watts, the Argonaut II is worth considering. If, however, you might want to run more power sometimes, you may find it limiting. ARRL publishes a collection of equipment reviews from *QST*. Pick up a copy if you're in the market for a used rig.[2]

Traditional QRP transceivers, like the Ten-Tec Argonaut series or the Heathkit HW-7, HW-8 and HW-9 are popular because they were designed especially for QRP operation. In fact, they have what amounts to a cult following among QRPers. If you need a smaller rig to take on trips, an Argonaut 509 or 515, or a Heathkit HW-9 may be a good choice. If you're going to operate from home however, why not just use your regular transceiver? You're comfortable with it, it's paid for and it is usually fairly simple to reduce its output to 5 watts or less.

First, you need an *accurate* QRP wattmeter, or some other

Figure 1—A low-power wattmeter is handy for adjusting your transmitter for 5-watts output.

means of measuring output power (Fig 1). You can use an oscilloscope or RF voltmeter to measure the output voltage across a 50-ohm load, and use Ohm's Law to calculate the power. Needless to say, a real wattmeter is much easier to use. Try to adjust the output power to 5 watts with the controls you normally use to adjust output. If you can't get the output power low enough, consult your owner's manual. There may be an internal adjustment that allows you to further reduce output.

ADJUSTING ALC VOLTAGE TO REDUCE POWER

If not, or if the idea of opening the box scares you, there are safe alternatives. If the transceiver (or transmitter) has an ALC input, you have a simple means of adjusting power output. The ALC input is meant to be connected to a power amplifier alc output, to prevent excessive drive which results in a distorted signal. Applying a positive bias to the transceiver/transmitter ALC input reduces output power. A positive voltage in the range of 5 to 12 volts is usually adequate. In no case should you exceed 12 volts. Fig 2 is a schematic diagram of a suitable circuit.

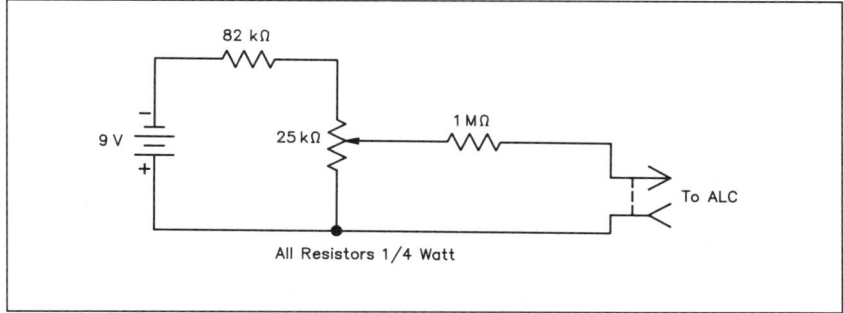

Figure 2—Schematic diagram of the ALC bias supply.

OTHER WAYS TO REDUCE POWER

Older transmitters, designed for CW and AM operation, don't have ALC inputs. Reducing the final-amplifier drive and retuning the final-amplifier plate circuit (we're talking about vacuum tube transmitters here) may not allow you to reduce power output to 5 watts. If you're comfortable inside such a transmitter, you probably already know how to reduce final-amplifier screen voltage, or even bypass the final amplifier altogether. You may be able to get 3 or more watts out of the driver tube. If you don't feel comfortable inside such a transmitter, I'd rather not describe such modifications. The voltages inside vacuum-tube transmitters are lethal. If you don't have experience doing such a modification, I'd prefer you tried another solution.

Another way to reduce power delivered to the antenna is to place an attenuator at the transmitter output (Fig. 3). Let's say your transmitter output can be reduced to 25 watts. The formula for gain or loss in decibels is:

$dB = 10 \log P2/P1$

Applying this equation to our case gives:

$dB = 10 \log 25/5 = 10 \log 5 = -6.9897 \, dB$

LETHAL VOLTAGES ARE PRESENT IN VACUUM TUBE RIGS!

A 7-dB attenuator will do fine. Tables of suitable resistance values for Pi- and T-network attenuators are given in the "Test Equipment and Measurements" chapter of *The ARRL Handbook for Radio Amateurs*. Note, however, that this attenuator must absorb 20 watts. The resistors used should not be wirewound, as they will look like coils at radio frequencies, not resistors.

To obtain sufficient power-handling capacity with attenuators built from 2-watt non-inductive resistors (carbon-film resistors will be okay up to 30 MHz), use more than one attenuator. For example, use a 3-dB attenuator that can dissipate 12.5 watts, followed by a 4-dB attenuator that can dissipate 7.5 watts. You can connect resistors in series and parallel to reach the correct resistance and wattage.

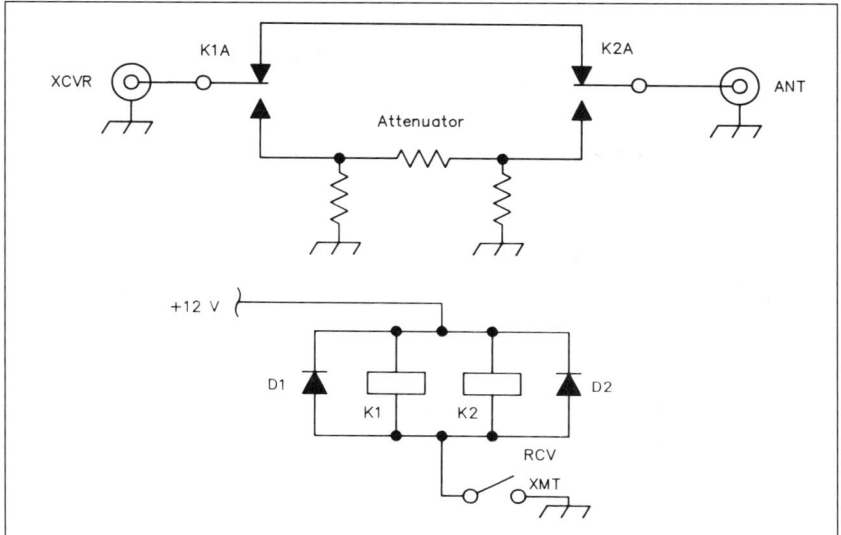

Figure 3—How to connect an attenuator to the output of a transceiver.
D1, D2—1N4001 or equiv
K1, K2—High-speed relays, Radio Shack 275-248

RECEIVERS FOR QRP

Running low power means sometimes it takes longer to work a station. Most amateur operating requires more listening than transmitting anyway, and you'll be doing even more listening than the average ham. Obviously, your receiver should be pleasant to listen to and easy to use. If you have a modern, solid-state transceiver, you may want to upgrade the I-F filtering, adding additional or narrower filters. An audio filter is another simple and inexpensive way to improve receiver performance.

If you plan to work other QRP stations, either in QRP contests or to increase your QRP Worked All States and DXCC totals, you need the best receiver you can afford. The ability to copy a weak signal when other, stronger signals are almost on the same frequency is probably the most-important receiver factor. You may think that high sensitivity is needed to hear weak signals, but the opposite is true. Atmospheric noise is usually strong enough that exceptional receiver sensitivity provides no benefits.

On 15, 12 and 10 meters, atmospheric noise may be low enough that additional receiver sensitivity can be useful, but only if it can be turned off when not needed. Many transceivers have built-in receiver preamplifiers *and* attenuators. When background noise is low and there is no adjacent interference, you can turn on the preamplifier to improve your ability to hear weak signals.

Sometimes the interference is so bad you have to use the attenuator! The attenuator reduces all signals by perhaps 10 or 20 dB, and may make weak signals unreadable. You need the attenuator when a very strong adjacent signal overloads the receiver, making it impossible to copy the desired station anyway. If you can turn on the attenuator and turn up the AF gain control to make up the difference, you may be able to pull the station out of the crud. If only *everyone* ran QRP!

STATION ACCESSORIES

Assuming you've been on the air at some time, you probably have a microphone, keyer and paddle. Do you have to pick up the microphone every time you speak, or lean toward it? Then you should buy a combination microphone-headphone boomset, like telephone operators wear. In a long contest, every step you take to reduce your fatigue is worthwhile. We're here to have fun after all!

You'll find most of your success in QRP comes from operating CW. You've heard all the arguments before, and you'll hear them again in a later chapter. For now, you should have at least a memory keyer and a good paddle. The memory keyer is not only useful for sending contest exchanges, but can save you a great deal of effort when you're calling a DX station in a pileup. Everyone who sends CW manually eventually makes a mistake. You don't want to make a mistake the *one time* the DX station hears you. Sending accurately requires practice. Don't ignore the paddle all year, then expect to be proficient when contest season rolls around.

A Computer in a QRP Station?

If you're serious about contesting, consider using a computer to do your logging. In fact, with the right computer and software, all you have to do is copy the other station's exchange. The best

contest-logging software is designed for use on IBM-PC-compatible computers. While any compatible will work, having a hard-disk drive is a real asset. General-purpose logging programs can also be used during contests, but they're slower and harder to use. What's worse, they don't produce logs in a format acceptable to most contest sponsors, including ARRL. Contest-logging software is advertised in *QST* and the *National Contest Journal*.

A personal computer is a worthwhile accessory, even when you aren't in a contest. Other programs are available to help you keep track of countries worked and confirmed, and forecast propagation conditions. If you're serious about DX and contesting, you'll probably want to connect to your local DX PacketCluster network (see Chapter 3). Your computer will serve nicely as a packet terminal.

A QRP station is usually simpler than one capable of higher power. For example, you won't have a power amplifier that requires a separate 240-volt power line and takes up space. In fact, your station may be small enough to pack in a suitcase, which presents all sorts of opportunities not available to those who run high power.

Notes
[1]D. Newkirk, WJ1Z, "The Ten-Tec Argonaut II HF Transceiver," *QST* January 1992, p 77.
[2]R. Schetgen, KU7G, Ed., *The ARRL Radio Buyer's Sourcebook*, ARRL, 1991.

CHAPTER 3
Operating Techniques

Succeeding against signals 10 times louder than yours requires operating skills the average ham never learns. Where a 100- or 1500-watt station can often plow through a pile-up with brute force, QRP requires some finesse. There's no magic involved really, and any DXer or contester would benefit from these skills. For now though, we'll keep them to ourselves!

DX TECHNIQUES

Some hams like to tune the bands, looking for pile-ups. They claim it saves them time when they want to work some choice DX. Busting pile-ups is one thing when you're running lots of power, but quite another when you're running 5 watts. If you can find the DX station *before* the mob does, you can probably make the contact and be on your way very quickly.

Rare DX stations are not often loud, nor are they very active. If they spent many hours putting out big signals, they wouldn't be rare, would they? If you want to work DX you have to look for the weak signals. A large percentage of hams who call themselves DXers will tune right past a weak signal. If the copy isn't easy they don't bother listening. This attitude is rough on QRPers! So, if we expect others to dig us out, we'd better be prepared to do some digging ourselves.

We avoid pile-ups as much as possible, because it's harder for a QRP signal to break a pile-up, but there are other good reasons.

Pile-ups draw attention. Pile-ups usually mean the station has been "spotted" on PacketClusters. Our high-power friends need to flex those kilowatts now and then, to justify the monthly payments. Meanwhile, a few kilohertz down the band, we've got some equally good DX all to ourselves! (Now you know why we don't want them to know our secrets.)

Of course, the day will come when you don't *want* to look elsewhere for DX. You *want* the DX in the pile-up! Think it over. Is this a DXpedition that's likely to be active for some time yet? Will they be begging for QSOs next week? If so, give yourself a break. Come back next week. Or, if the station operates on 12 or 17 meters, try to work it there. Pileups on these bands *are getting bigger*, but they're still smaller than those on 20, 15 or 10.

If there's no way out of joining the pile-up (a short-term operation, for example), you'll have to be crafty to get a QSO. Your best chance is probably on CW, when the station is working "split." *The DXCC Companion* has several good tips for working DX in pile-ups[1]. The bottom line is, to make contact you have to be transmitting on the frequency where the other operator is listening, when he or she is listening. This advice seems so simply logical, it's amazing more DXers don't know it or heed it!

Let's say the DX station is transmitting on 14025 kHz. You hear the operator call N1BKE. Switch VFOs and tune up in frequency (most stations listen above their transmitting frequency; a small but significant number listen below). You're trying to find N1BKE as he sends his half of the exchange. You don't have much time, as DX exchanges are short, and you'll be distracted (and amazed!) by the other stations calling at the same time.

Although it may take several minutes to find a station actually *working* the DX, when you do you're in the race. You can try "tail ending" (sending your call once as the other station signs), but this technique works only rarely, especially for QRP stations. You have to be nearly as loud as the other station to be noticed by the DX. Switch VFOs and listen to the DX station again.

At least you know the area of the pile-up where the DX is listening. When the DX acknowledges the contact ("TU" on CW, "thanks" or "QRZ" on phone), switch to the other VFO. The *first* time you call, move a little higher in frequency than the pack calling on the frequency where the last station got through. You want to be clear of the mess. Send your call once, a little slower than the pack (about 15 WPM is good). Listen to the DX station again.

If you didn't get through, you have to find out who did. The DX station may have tuned an exact kHz higher after the last QSO. Now you know what to do next time. Of course, the DX won't keep tuning higher forever. Sooner or later, he or she will have to return to the starting frequency, say 14030 kHz. Some operators tune lower in 1-kHz increments, some jump right back to the starting point.

DON'T CALL ON TOP OF DX!

Other DX operators prefer to pick stations off the top and bottom of the pile-up. Tune higher or lower until the number of stations thins out. Then go out a little farther and call there. Unless you hear other stations on the DX station's frequency getting through, don't call on top of the DX. No matter how deserving you feel, with your environmentally sound 5-watt station, you're just going to enrage the other DXers trying to copy the DX station.

Given enough time and some luck besides, you'll eventually be heard by the DX station, *if* you call at the right places and times. Let the other stations waste time and power blazing away on one frequency. Yours is a DXer's station, not a beacon.[2]

Pile-ups don't sound like much fun, do they? They are, if you get through! It's a great feeling to get through, when more powerful (but less skillful) stations don't. However, you are forgiven for spending your time looking for stations you can work, rather than beating your head against the wall of a pile-up.

DXing The QRP Calling Frequencies

By general agreement, certain frequencies on each amateur band from 160 to 10 meters are where QRPers look for other QRPers. Table 1 lists those frequencies. During band openings you can usually stir up some activity by calling CQ DX on or near a calling frequency. Some home-built QRP transmitters are crystal controlled on these frequencies, with provisions for swings of perhaps 10 kHz.

Most of the DX you'll work though, will not be operating QRP. A great many DX stations operate in the Extra Class CW subbands, which is why most competitive QRPers have that license. While quiet most of the time, the calling frequencies do get busy during QRP contests.

PacketCluster

The PacketCluster system for DX and contest spotting was a major development in Amateur Radio competition.[3] DX stations

TABLE 1
QRP Calling Frequencies

Band	Phone[1]	CW	Novice
160	—	1810	—
80	3895	3560	3710
40	7285	7040[2]	7110
30	—	10106	—
20	14285	14060	—
15	21385	21060	—
12	24959	24900	—
10	28885	28060	—
6	50885		

[1]There is very little QRP activity on the SSB calling frequencies.
[2]In Europe, 7030 kHz is the 40-meter CW calling frequency.

Operating Techniques 3-5

can always tell when they've been spotted on a local node: suddenly, every station calling is from the same area! PacketCluster is a good way to keep in touch with other DXers and contesters in your area. Bulletins and messages carry information on upcoming DXpeditions and other matters of interest to us. For the QRP DXer however, the actual spots are not always the blessing they might seem. Once you know about the station that was spotted, so do hundreds of other hams, most of whom are running more power than you. That's why it's important to spend time tuning the bands, finding the good stuff first. After you have the DX in your log, you can be the one to spot it! Go ahead and gloat that you worked it with 5 watts, 2 watts or whatever power you run. Wait until the pile-up gets huge, then go dig out another good one. Now *that's* DXing!

CONTESTING

Thinking "weak" is probably the biggest mistake made by QRP contesters. Some QRP contesters are self conscious about their low power. They repeat their exchange two or more times, because they think the other station can't possibly copy their QRP signal. Spend some time watching your S meter and you'll see the average contest signal is *very* strong, and QRP signals are *strong*! If the other operator misses part of your exchange he or she will tell you. Send it once, but get it right the first time!

Yes, QRPers have to be more careful than block-buster stations. Other stations don't mind concentrating a little harder to dig us out of the QRM, but don't torture them with mistakes and repeats. A contester who's trying to "run" other stations (letting them call her, and seeking a large number of contacts per hour) doesn't want another station to take her frequency. To keep the frequency, she can't let much time elapse between transmissions. Fumble around and repeat your exchange and you may find the other station just starts transmitting on top of you! This tactic isn't meant to hurt your feelings; it's a matter of survival.

If you don't use a computer logging program that sends CW,

invest in a memory keyer. Carefully load your call and the contest exchange into the memories, so you won't mess up under pressure. Sloppy operating, not weak signals, gives QRP a bad name.

In addition to being accurate and copiable, you should be transmitting on the right frequency. During popular contests, stations may be operating a kHz or less apart across the band. You must be able to zero beat the station you're calling or you're wasting time and causing QRM. The easiest way to learn how to zero beat your rig to another station is to set up a second receiver. Tune in the same station on both rigs, then (preferably into a dummy load) transmit to see how close you are to zero beat. You have to learn at what received-signal pitch your transmitter is tuned to the same frequency as the other station.

This pitch may be uncomfortably high or low. If so, consult the manual and adjust it for your comfort. If you can't adjust the *transmitter offset* inside your rig, you may be able to permanently set the receiver-incremental tuning (RIT) to compensate. However you do it, it's important to call the other station on his or her frequency. You'll make many more contest QSOs that way.

Other contesting techniques for QRPers vary, depending on the contest. Let's look at several popular contests and see what techniques work best.

DX Contests

The most-popular DX contests are the ARRL DX Tests and the *CQ* World-Wide (WW) DX Contests. Both have "QRP" categories, which means we're competing against each other, not against the high-power gang.

Multipliers in the ARRL DX Test are countries other than the US and Canada. In the *CQ* WW test, you have to work the US and Canada once on each band; other than the one contact per band, contacts with the US and Canada don't count. *CQ* WW multipliers are DXCC countries and *CQ* zones. *The ARRL Operating Manual* contains maps showing the boundaries for the 40 different zones.

Operating Techniques 3-7

CQ CONTEST!

After finishing the yard work and making a trip to the hardware store, you finally get a chance to sit down at the radio on Saturday afternoon. All you hear is "CQ TEST CQ TEST." Now what? Well, you have several options. You can operate on another mode. You can check out 30, 17 and 12 meters, where contest operation doesn't take place. You can check out college football on the tube. In other words, you can ignore all those other hams having a good time. Next time you hear a contest, give it a chance. There must be *some* reason why so many hams are such avid contesters. Let's look at the pros and cons.

Yes, some hams absolutely despise contests. They're entitled to their opinion of course, but let's take a closer look at this popular activity. Contests generate loads of activity on otherwise pretty uncrowded bands. You can learn more about how band openings shift and fade in one weekend of contesting than in years of "normal" operating. You can also work more DX in a weekend than many noncontesters work in a lifetime!

"Ah," some say, "contest contacts are shallow, and meaningless outside the context of the contest rules." Sure. And the average Amateur Radio QSO is loaded with social and intellectual content, right? At the Woodstock Music and Art Festival in 1969, I probably didn't talk to more than a half dozen people, other than to say "hi." Yet there was the great feeling of sharing an experience with

(*CQ* zones are different than IARU zones, which are used in the IARU World Championship contests.)

Your goal in a DX contest is to work as many multipliers as possible *on each band*, and also to work as many stations as possible. You should divide your time between multiplier hunting and trying to build up your contact total. Knowing when a part of the world from which you need multipliers should be workable on each band is helpful.

thousands of other like-minded people. Contesting is like that. You may not stop to talk to a friend more than once or twice during the contest, but, if you stop to think about it, you're sharing the fun and excitement with thousands of other hams at the same time.

"Did you say excitement?" someone asks. Yes, contesting is exciting. You never know what the next turn of the dial or the next CQ will bring. Hearing all six continents at the same time, within a few kHz, brings back the excitement we felt as beginners, when *every* contact was an adventure.

Contesters *are* social! Do your club meetings attract over 100 attendees? Those of the Yankee Clipper Contest Club (YCCC) do! Every meeting includes several presentations on antennas, propagation, computerized logging and other subjects of interest to active hams. There's always time for some good-old eyeball QSOs. You just don't find that kind of camaraderie in most general-interest clubs. The other club members will try their best to beat you in the next contest, but in the meantime they'll gladly help you put up your new tower or antenna.

So, the next time you stumble onto a contest, don't complain that it's interfering with your fun. Instead, look up the rules in *QST* and make the fun of contesting part of your Amateur Radio life. — *Jim Kearman, KR1S*

Although you'll spend a fair amount of time calling other stations, don't overlook the chance to call CQ. DX QRP stations in particular will be looking for other stations to call. Your best success will probably be on the highest-frequency bands open, but don't ignore the lower bands. Even if you aren't loud enough on the lower bands to successfully call CQ for long periods, you can easily pick up multipliers. After you've scanned 20 for multipliers, for example, move back up to 15 and 10 and try to build up contacts.

ARRL Sweepstakes

Contesters have a love-hate relationship with this event. Some dislike the long exchange, others like the challenge. Sweepstakes multipliers cover the US, Canada, Puerto Rico and the US Virgin Islands. The Pacific section, another multiplier, includes Hawaii and the US territories in the Pacific. There is a separate entry category for QRP stations.

Unlike DX contests, you can only get credit for working a station once, rather than once per band. This rule is an advantage for QRPers. You can work the most-distant sections on the bands that favor those distances, and not be penalized for not working them on the other bands.

With fewer than 80 multipliers available (the number may change so it isn't given here!), it makes sense to spend some time looking for them. States like North Dakota and Mississippi may be represented by a single entrant, so you'll want to find that station when it's on.

Forty meters will probably provide most of your contacts, but you'll have to use other bands to work sections farther away. Coast-to-coast two-way QRP contacts on 40 meters are not unusual in Sweepstakes, and the QRP category is very popular. Sweepstakes is an excellent contest for increasing your QRP Worked All States tally. It pays to call CQ when you can find a clear spot, as some rare sections can only be worked this way.

At night though, 40 meters is a jumble of contesters, foreign broadcast stations and SSB stations in Canada and the Caribbean. This is no place for a QRP station to call CQ! You can work other stations by calling them, but do your 40-meter CQing during the daytime, when interference is less dense.

This strategy works on other bands as well. The high-power stations, which mostly call CQ, need to be on the most-active band. You also want to be on the band that brings you the greatest number of contacts per hour, but that isn't necessarily the busiest band.

QRP Contests

QRP Amateur Radio Club International (QRP ARCI) sponsors several contests for QRPers. Multipliers are US states, Canadian provinces and DXCC countries (States, Provinces, Countries—SPC). QSOs with QRP ARCI members are worth 5 points, non-members outside North America are worth 4 points, but non-members in North America are worth only 2 points. A station can be worked once per band, as in other DX contests.

Unfortunately, there aren't enough QRP ARCI members worldwide to allow you to run up a big score by working only members. Most of your DX contacts will be with non members. The exchange you must receive from a non member is: Signal report, S/P/C and *power output*. You can infer the country from the station's prefix, but getting the power information requires some work. You'll probably have to rag chew with the station, and you may have to ask: "RIG?" or "POWER?" to get that last piece of information. At 4 points per QSO, and especially if you're getting a new multiplier, it's worth rag chewing for a couple of minutes. North American stations who aren't in the contest, on the other hand, are not worth the trouble *unless* they provide a new state or province.

Field Day

While not considered a contest, Field Day is custom-made for QRP operation. It's exciting to operate QRP from home, but the thrill of making contacts from out in the boondocks, using wires tossed up in the trees and batteries for power is indescribable.

The Field Day multiplier is based on your power level, and you only work stations once, regardless of band. If you operate CW and SSB, you can work a station once on each mode. CW QSOs are worth 2 points, versus 1 point for SSB. Obviously, most competitive Field Day operations are CW only.

Field Day antenna schemes depend on how large a crew you

Figure 1—Field Day was made for QRP! WØBXS operates W6LFB/6 on Field Day 1991.

can put together, and where you're located. As in Sweepstakes, 40 meters generates the most contacts, but it's a good idea to put out a good signal on the other bands as well. Quads or delta loops hung from trees work well on the high bands (14 MHz and up). If you live on either coast, you can add a director element to increase the antenna gain. In the central states, a bidirectional loop will keep you going for hours.

Contacts on 30, 17 and 12 meters are acceptable on Field Day, as it is a test of operational ability. Activity on these bands is light, however, so don't worry if you can't work them.

As Field Day is not a contest, award certificates aren't issued to the "winning" stations. Regardless, Field Day is a great way to spend the weekend afield with friends, having fun with QRP (Fig 1).

Computer Contest Logging

Winning a contest in the higher-power categories requires computer logging. With a good computer-logging program you can tell instantly if you need a station or a multiplier, know if the station

SOME COMPUTERS GENERATE ALMOST AS MUCH RF AS YOUR QRP TRANSMITTER!

you're about to work is a dupe, and be able to see at a glance how your score is progressing. Say, we QRPers could use the same kind of help! Why waste time calling a station when you don't need the multiplier? And who wants to spend hours after the contest checking for dupes, or recopying a messy log? Yes, computer logging is for us, too (Fig 2).

There are several computer contest-logging programs available, but a few have advantages worth knowing about. All the best programs are made to run on IBM-PC-compatible computers. They're advertised in the *National Contest Journal, QST* and *CQ*.

If you operate CW, buy a program with a built-in keyer. You

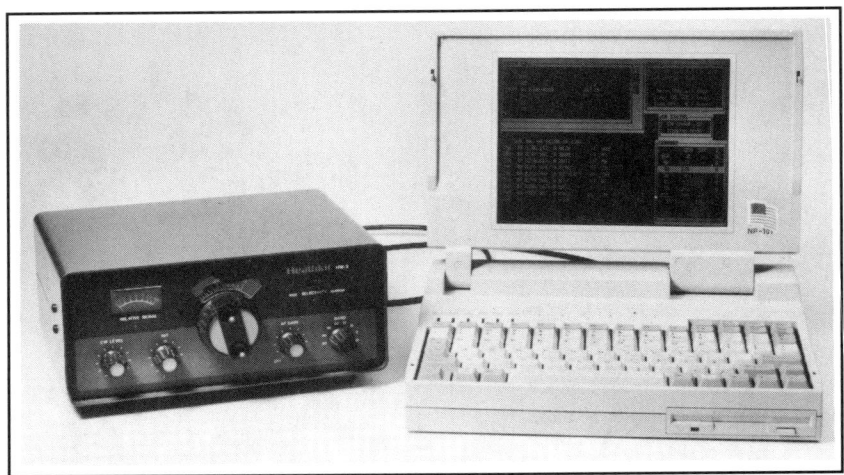

Figure 2—Computer logging takes the drudgery out of contesting.

run a cable from the computer serial-data or printer port to your key jack. You can program the computer function keys to send all or part of your exchange, call CQ and so on. In a contest like Sweepstakes, where the exchange includes a serial number that increases with each QSO and the exchange itself is long, this feature is priceless!

Improvements in PC technology have sent thousands of perfectly healthy XT-model PCs into the used market, and new 80286- and 80386-based computers are more affordable than ever. Many computer-logging programs will run on floppy-disk-only systems, but new releases of these programs are too large to fit on a 360-kbyte disk. You should buy a computer with a hard-disk drive for contest logging.

Some early computers generate almost as much RF as your QRP transmitter, so it's a good idea to try it near your rig before you buy if you can. Some of the hash comes from the computer box itself, but the video monitor may also cause problems.

Once you've operated an entire contest without having to pick up a pencil or touch your paddle, you'll wonder how you ever got by without computer logging. When the contest is over you print out the log and mail it in. For ARRL contests, you don't even have to do that! ARRL accepts floppy disk entries in the proper format, or you can send your log over the telephone lines to the ARRL computer bulletin board. Don't forget to include a file with the summary sheet data. If you have one of the popular file-compression programs, like PKZIP, it's okay to ZIP the two files before you send them.

Notes
[1] J. Kearman, KR1S, *The DXCC Companion*, ARRL, 1990.
[2] J. Kearman, KR1S, "Pagalu Safari," *QST*, May 1990, pp 42-43.
[3] J. Cain, K1TN, "Packet with a Purpose," *QST*, Aug 1990, pp 54-57.

CHAPTER 4
A Propagation Primer

Large investments in antennas, feed lines, and towers can be wasted if you don't understand what happens to your signal after it disappears over the horizon. Successful QRPers know when to operate and where to point their beams for best results.

SUNSPOTS: THE COOPERATIVE SUN

Sunspots and solar flares are two features of the sun of particular interest to the QRPer. Both directly affect the ionosphere and in turn, HF communications.

Sunspots are the most obvious surface feature of the sun. Very large groups may be visible without magnification. (NEVER look directly at the sun with the naked eye, binoculars, or a telescope. Direct exposure of the retina can result in blindness. Instead, use a filter known to be safe—one that passes no more than 0.01% of the impinging light across the spectrum from infrared through ultraviolet.) The number of sunspots varies in an approximately 11-year cycle.

Help from the Ionosphere

Long-distance HF communications are possible because of the ionosphere, that region of the atmosphere which lies between 60 and 200 miles above the earth. When the number of sunspots is high, radiation from the sun that affects the ionosphere is also high. Ultraviolet light and X-rays striking this region create ions, which are capable of refracting radio waves.

Radio waves are not reflected, but rather bent, or *refracted* through the various layers of the ionosphere. Waves that leave the antenna at lower angles above the horizon require less refraction to return to earth. Low angles of refraction also mean less energy lost and greater distances achieved. For these reasons, QRP operators strive for low radiation angles from their antennas.

An exception to this rule of thumb is when you want to have good local coverage. For example, during the ARRL Sweepstakes contests, you'll use 80/75 and 40 meters mostly for contacts out to 1000 miles. Most contacts on these bands will be at distances under 500 miles, where high radiation angles are desirable. Chapter 5 discusses antennas for high and low angles of radiation.

One useful indicator of solar activity is the solar flux value. Measurements are made daily at several frequencies. Of interest to amateurs is the value at 2800 MHz. Radio station WWV in Colorado broadcasts solar flux and other related information at 18 minutes past the hour; WWVH in Hawaii broadcasts the same information at 45 minutes past the hour. See the sidebar for more information on these broadcasts.

Regions in the Ionosphere

As the atmosphere becomes thinner at higher altitudes, layers or regions form in the ionosphere. Ionospheric regions have different effects on signals propagating through them.

D Region

The lowest ionospheric region is the D region. Its height is about 45 miles. The amount of D-region ionization is proportional to the angle of the sun, and is greatest when the sun is directly overhead. The D region can refract radio waves at very low frequencies (VLF), but it absorbs radio energy at higher frequencies.

Absorption is greatest in the MF and lower end of the HF portions of the radio spectrum (1 to 10 MHz). As a result, propagation on the low bands (160-40 meters) is normally limited to regional coverage (under 300 miles) during daylight hours.

At higher operating frequencies, however, D-region absorption drops. You should try to operate on the highest frequency band open to the place you want to work; lower absorption means your signal will be stronger. The D region quickly disappears as the sun sets, making the low bands usable for DX just after sunset. During times of moderate to high numbers of sunspots, these bands may stay open all night, until just after sunrise.

E Region

The lowest useful region of the ionosphere for HF propagation is the E region. It is approximately 70 miles above the earth's surface. E-region propagation sometimes assists long-distance communications over daylight paths. The E region disappears soon after dark.

Sporadic E Propagation

Small, scattered clouds of intense ionization can be formed at the heights of the E region. These clouds allow what's called sporadic-E or E_s propagation. These ionized patches vary, both in size and intensity, and generally move in a northwesterly direction. They are most prevalent in the late spring through early summer in the middle latitudes.

Sporadic-E clouds are probably formed by wind shear at high altitude, often associated with large thunderstorm systems. If the weather forecast is for thunderstorms hundreds of miles from your location, there's a good chance for sporadic-E openings.

Sporadic-E openings are short and erratic, but signal levels are often high. The skip distance for a typical opening is hundreds of miles, but multiple hops yielding distances up to 2500 miles aren't uncommon. From the continental US, sporadic E propagation is a good way to pick up new states and even some Caribbean and Central American countries on 15, 12 and 10 meters.

With the target area of propagation shifting as rapidly as signal levels, E_s QSOs can quickly become contest-style exchanges, with little or no ragchewing. When solar activity is high, the effects of sporadic E are often masked by the excellent F-region propagation.

F Region

The region of the ionosphere that permits most HF communications is the F region. During daylight hours, this region splits into two layers or regions: the lower F1 and upper F2 regions. The F2 region is primarily responsible for refraction of HF radio waves. Because of the low density of the atmosphere at this high altitude (175 miles), ions take a long time to recombine with free electrons. For this reason, when solar activity is high, the F region often lasts all night. Long-distance contacts on the high bands (14 MHz and above) are often possible all night. Because the D region has disappeared, these are prime hours for QRP DXing.

For a given location the F region reaches greatest ionization in the early afternoon; ionization is lowest just before sunrise. When there are many sunspots, the F2 region often stays ionized all night. When sunspots are few, the F-region maximum usable frequency (MUF) may drop below 3 MHz, making all bands except 160 meters useless for DXing.

The F-region MUF pretty much determines the highest frequency you can use to work into a particular area. Because D-region absorption decreases as you go higher in frequency, it pays to operate as close to the MUF as possible. This is especially true when you're running low power. More powerful stations can still put out readable signals on 20 meters when the MUF is over 30 MHz. Remember, their output power may be 300 times yours, and they may have bigger antennas besides!

Ionospheric Scatter

When the nighttime MUF drops below 3 MHz during a major contest, the only way to work over the horizon is by scattering your signal in the thin, remaining ionosphere. Signals propagated by scatter sound hollow and are often weak. If you're a QRP operator, you probably won't get many answers to your scatter CQs, but you'll get more sleep than the high-power operators!

SOLAR FLARES: THE UNCOOPERATIVE SUN

Eruptions from the sun's surface, called solar flares, sometimes increase absorption in the ionosphere. Flares may also cause geomagnetic storms. A geomagnetic storm is a worldwide disruption of the earth's magnetic field.

Very often, D-region absorption increases as the result of a flare. The result is a radio "blackout" on certain paths. HF communications that cross the polar regions grind to a halt for several hours, until the effects of the flare dissipate. Contacts along east-west lines are still possible at reduced signal levels, but the best bet is to point your antenna south (from the northern hemisphere) and try to work stations nearer the equator, where the effects are less severe.

Not every large flare causes a disruption in the ionosphere. The position of the flare on the surface of the sun determines if the particles it ejects will interact with the earth's magnetosphere. One of the largest solar flares ever recorded, in August 1989, was located on the west limb of the sun. Since it was pointed away from the earth, its stream of charged particles was only slightly influenced by the earth's magnetic field. As a result, it only slightly disturbed the ionosphere.

Geomagnetic storms sometimes follow solar flares. A storm may begin from 16 to 36 hours after the flare. When the earth's geomagnetic field is at storm levels (as reported on WWV and WWVH), the F region is often greatly disturbed, to the point that it seems to disappear. Propagation on the high bands is often very poor, but the low bands may still be usable after dark. If you're DXing, disturbed conditions can work for you. Many less-dedicated DXers will turn on the radio, see that conditions are pretty bad and go do something else. With your competition off watching TV, the occasional DX that gets through on 80 and 40 meters is yours for the asking!

HOW FAR CAN YOUR SIGNAL GO?

The distance at which a radio signal returns to the earth depends on the radiation angle of the transmitted signal and the height of the ionospheric region that refracts it. As region height increases, so does "skip" distance. The maximum distance that can be covered in a single hop is about 1250 miles when refraction is via the E region. When the wave is refracted by the F region, this distance increases to a maximum of 2500 miles. However, these estimates assume a near zero-degree angle of radiation, which is nearly impossible to achieve with practical antennas.

When a radio wave returns to earth, it is scattered, which may cause multiple-hop propagation. This process may be repeated several times, allowing the signal to travel all the way around the world. However, ionospheric attenuation and ground reflection losses take their toll with each hop. Multiple-hop transmissions lose signal strength with each refraction and reflection. A lower radiation angle reduces the number of hops, which makes your signal stronger.

PROPAGATION FORECASTING

In spite of the many complexities of propagation, there is a method of determining communications possibilities in advance. Computer programs have been developed that provide reasonably accurate forecasts of band openings. Armed with these predictions, you can select a band to use at any time of any day during the year.

The better programs give an expected signal-strength level for different frequencies along the path between two points. You can

```
MINIPROP (TM) SHORT-PATH PREDICTIONS      06-16-1992       Path Length  :   7790 km
Sunspot Number :  76.8      Flux : 121.0     F Hops : 3    Radiation Angle :  7 deg
TERMINAL A :  47.60 N    122.60 W    Port Orchard, WA      Bearing to B : 204.5 deg
TERMINAL B :  19.00 S    147.00 W    French Polynesia      Bearing to A :  17.2 deg
Terminal A Sunrise/Set : 1218/0404 UTC    Terminal B Sunrise/Set : 1623/0314 UTC

                     ----------------- SIGNAL LEVELS in dB ABOVE 0.5 uV -----------------
 UTC   MUF    3.5 MHz   7.0 MHz   14.0 MHz   21.0 MHz   28.0 MHz
 0000  20.9                          7 A       20 C
 0030  20.8                          9 A       20 C
 0100  20.9                         11 A       21 C
 0130  21.0                         19 A       22 B
 0200  21.1                         21 A       23 B
 0230  21.1             -2 A        23 A       24 A      23 D
 0300  21.1              6 A        26 A       26 B      24 D
 0330  21.1             19 A        28 A       27 B      25 D
 0400  21.0    11 A     27 A        31 A       28 B      2/
 0430  20.9    32 A      9           32 A       29 C
 0500  22.1    //                   33 A       29 B
 0530  22.^                         33 A       29 r
 06^^                                ^          2'

 2030  20.8                                    17 C
 2100  20.9                                    17 C
 2130  20.9                         _ A        17 C
 2200  21.0                          3 A       17 C
 2230  21.0                          3 A       18 C
 2300  21.0                          4 A       18 C
 2330  20.9                          6 A       19 C

 Signal levels not shown if below -10 dB or if predicted availability is zero.
 Availabilities  A: 75 - 100%   B: 50 - 75%   C: 25 - 50%   D: 1 - 25%
```

Figure 1—MINIPROP is a propagation forecasting program for IBM-PC-compatible computers. Working from latitude, longitude, date and solar-flux data supplied by you, MINIPROP predicts signal levels over a 24-hour period.

adjust the program parameters to match your antenna height and gain, transmitter power and receiver sensitivity. This feature is especially helpful for the QRP operator. The programs are usually supplied with the parameters set for higher power, say 100 watts. By adjusting the parameters for your station, you can easily find out how loud you'll be at the other end of the path you want to work.

Propagation-forecasting programs are also useful for predicting openings you might not suspect, like late-night openings on the high bands when you thought they closed down at sunset. Fig 1 shows a sample screen from the propagation forecasting program MINIPROP.[1]

PROPAGATION PATHS

Radio waves normally follow the most direct path between the transmitter and receiver. This is the *great circle*, the shortest distance between the points. There are actually two paths between the transmitter and receiver. Unless the two stations are exactly opposite each other on the earth's surface (antipodal), one path is shorter than the other. Most of the time, QRP operators use the short path. Sometimes however, the short path is not open but the long path may produce contacts.

Long Path

Most DXers are aware that long-path openings may exist at any time to any part of the world. This is particularly true on 80 through 20 meters, although openings may occur on any band. During periods of high sunspot numbers, even 10-meter long-path openings are common.

A number of conditions must be satisfied before the QRPer can hope to find or use a long-path opening. First, the long path will seldom be better than the short path if the DX station is less than 6000 miles away. Second, the long path must be mostly through the nighttime hemisphere, to minimize D-region absorption. Third, solar activity must be high enough that the MUF does not drop below the band in use during the hours of darkness. Finally, the geomagnetic field must be relatively quiet (K index—from WWV/WWVH—less than 4).

From the northern hemisphere, the best long-path openings are to the south or southwest in the morning, and to the north or northwest in the evening. Contacts in excess of 20,000 miles have been made, especially when both stations are located along the sunrise-sunset terminator.

Gray-Line Propagation

The line dividing the region of the earth in daylight from that region in darkness is called the *terminator*. It is not a distinct division, but a gray, twilight band stretching completely around the earth. Any point along this *gray line* is at either sunrise or sunset.

Being lower in altitude and more dense than the F region, the

THE EARTH'S AXIS IS INCLINED ABOUT 23.5 DEGREES

D region is not ionized as quickly in the morning. Thus, the F-region MUF may rise quickly over the gray line as the earth rotates, while D-region absorption is still low. At sunset the D region quickly disappears, but the F-region MUF drops gradually. It's easy to see why signals propagated along the gray line are often surprisingly strong. This feature of gray-line propagation allows QRP stations to make long-path contacts over incredible distances.

The position of the terminator changes as the earth orbits the sun. The earth's axis is not perpendicular to the ecliptic (the plane of the earth's orbit around the sun), but is tilted about 23.5 degrees. Because of this tilt, the terminator swings back and forth about 47 degrees during the year. The northern and southern limits of this motion are defined by the Arctic and Antarctic Circles (23.5 degrees north and south latitudes).

A Propagation Primer 4-9

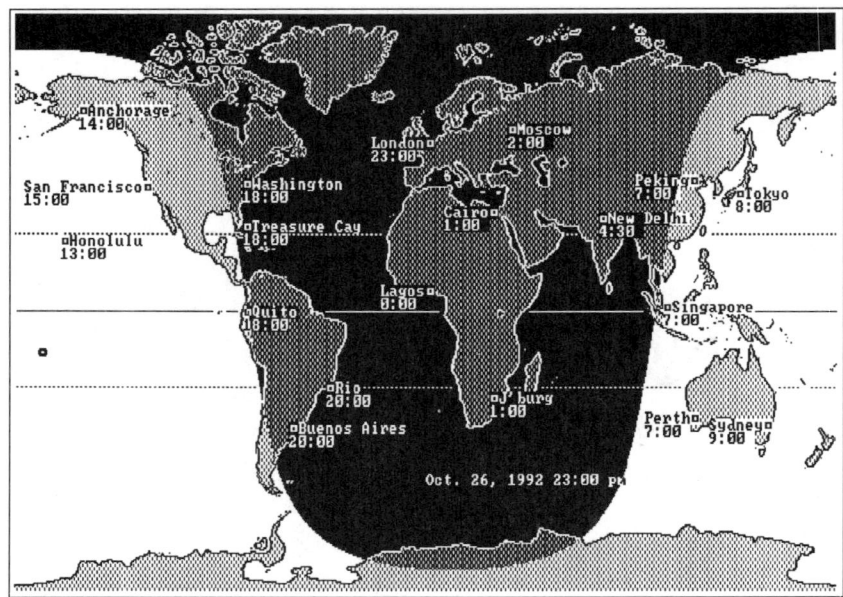

Figure 2—The *Gray Line Advantage* program depicts areas of the world in darkness and daylight. The *terminator* between the two areas is the gray-line boundary.

During the five or six sunspot-cycle years of low activity, gray-line propagation is most effective on 160, 80, and 40 meters. During years of high sunspot numbers, gray-line propagation is possible even on 10 meters.

Because the terminator extends completely around the earth, under the right conditions it is possible to work any area of the world *within this region*. Unfortunately, there are areas you'll never work on a gray line, because they never share a terminator with your location. This would only be possible if the earth's rotational axis were tilted 45 degrees to the plane of the ecliptic.

Two other factors affect gray-line DXing. Local time varies along the length of the terminator. More hams are active in the early evening than at dawn. Try to operate when the DX station is most likely to be active. To work someone halfway around the world means you're in front of the rig 30 minutes before local sunrise. In addition, remember that the direction of the terminator at your sunrise is not the same as its direction at local sunset.

A convenient method of figuring gray-line paths is through the use of *The DX Edge* slide rule or one of the computer programs such as MFJ Industries Gray Line Advantage and GEOCLOCK (GEOCLOCK is available for downloading from computer bulletin boards around the country) that shows areas of the earth in daylight and darkness. Fig 2 is an example display from Gray Line Advantage.

Crooked-Path Propagation

Let's say you want to work Europe on 10 meters during the Worked All Europe contest. When you point your beam at Europe you hear signals but they're weak. Yet, when you swing your beam a little south of Europe the signals get louder. For any of several reasons, radio signals don't always follow the great-circle path between two points (Fig 3).

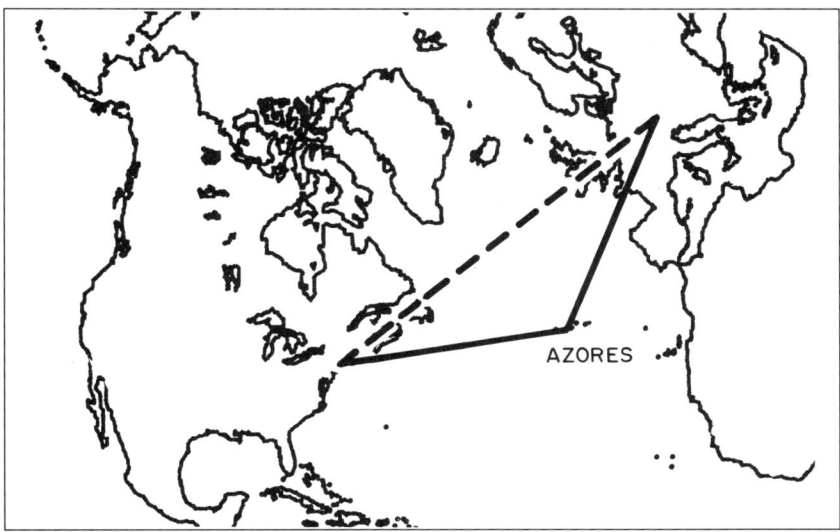

Figure 3—This map shows the direction of a typical crooked-path opening between the northeastern US and Europe. Propagation, on 10 meters for example, may not exist on the expected great-circle path. Europe and the northeastern US may, however, each have good 10-meter propagation to the Azores. Amateurs in both locations often can communicate best by pointing their beams at the Azores, rather than at each other.

Crooked-path propagation is one reason it is possible to work areas of the world on certain bands when propagation forecasts indicate there are no openings at that time.

A Propagation Primer   4-11

Crooked-path propagation is found on all bands, on short and long paths. Long paths are often crooked, so it pays to swing your beam off the expected path and see what happens to the signal you're copying.

PROPAGATION THROUGH THE YEAR

Much of the absorption and refractive characteristics of the ionosphere are directly related to the angle of incoming solar radiation. When the sun is low on the horizon, absorption characteristics change more rapidly than refractive characteristics. Because of the rapidly shifting properties of the ionosphere within the terminator region, the propagation of certain frequencies along the length of the terminator is extremely efficient.

Another factor affecting the ability of the ionosphere to refract radio signals is the seasonal variation in the angle of solar radiation. This is an annual cycle related to the season of the year. In March and September, the earth's rotational axis is at right angles to the sun. In June and December, the earth's axis is tilted approximately 23.5 degrees toward, or away, from the sun. This changing angle of radiation affects the amount of solar radiation that the ionosphere receives at various latitudes.

The reduced solar radiation of winter generates fewer ions in the ionosphere to refract radio waves. The longer nights give these ions more time to recombine, further reducing their concentration and reducing propagation possibilities. Thus, the 10- and 15-meter bands tend to die after early winter sunsets, even during periods of high solar activity.

As the days grow longer, the angle of incoming solar radiation is increased, which results in increasing levels of ionization. Thus strengthened, the ionosphere will support propagation on 10 and 15 meters beyond the local sunset.

However, the bands do not continue to improve as summer approaches. In the Northern Hemisphere they begin to deteriorate again after the vernal equinox (about June 21). While the increased radiation of summer generates more ions, it also heats up the atmosphere enough to cause some vertical mixing of its normally separated (stratified) regions. As lower regions mix with the upper, more ionized, regions, there is a reduction in the concentration of ions and free electrons. This dilution reduces the MUF and limits high-band propagation. The cycle begins to reverse as summer passes and fall approaches. All things being equal, the best high-band propagation occurs during the spring and fall months, around the equinoxes.

Propagation, and the many factors that affect it, are complex subjects. Reading is no substitute for experience. Time on the air is still required to learn the subtle characteristics of each band. Each band offers different opportunities for the QRP operator. Use your 5 watts as a tool to explore the current conditions. The smart QRP operator uses propagation knowledge to circumvent, not fight, the ionosphere.

The A index is a measure of activity for the earth's geomagnetic field within the past 24 hours. Values for the A index can range from zero to something in excess of 400, but values above 100 are very rare. When the A-index value is below 10 and the geomagnetic field is relatively stable, excellent communications may be expected. The absorption of radio energy by the ionosphere is low at this time, yielding strong signals over long-haul paths. If the circuit path lies over polar regions, The A-index value must be low or propagation will become nonexistent. Low-latitude paths or those crossing the equator are the least affected by a high A-index value.

The K index is similar to the A index except that it reflects more current conditions. This reading is derived eight times daily. Both WWV and WWVH geoalerts are changed every three hours to reflect these new values. The K index is a logarithmic scale rather than the linear scale of the A index. Thus, it is most sensitive at the low end of the scale, while high values change little in response to conditions. This figure is computed for Boulder, Colorado. A station to the north may experience higher readings, while a station

A Propagation Primer   4-13

to the south may have correspondingly lower readings.

Solar activity, as used in the geoalerts, relates to changing conditions that may adversely affect propagation on high-latitude circuits. This activity is described in terms ranging from "very low" through "moderate" to "very high."

Geomagnetic field conditions are described as "quiet," "unsettled," or "active." Quiet field conditions relate to an A index of 10 or less, while active describes an A index greater than 27. In addition to field conditions, geomagnetic storms will be included and described as minor or major. Geomagnetic field activity has the least effect on transequatorial paths, but an active geomagnetic field with major storm levels can close most HF bands to all but ground-wave propagation.

Another term used in these geoalerts is "stratwarm." It is a shorthand term for stratospheric warming, a seasonal condition existing in the polar regions. It is essentially a vertical mixing of the atmosphere which reduces the level of ionization in the ionosphere. These lower levels of ionization drop the MUF and reduce propagation opportunities on circuits that cross these regions. Stratwarm conditions are primarily a meteorological phenomenon, and are little affected by the current phase of the solar cycle.

For the A and K indices, the important thing is not so much the day-to-day values, but rather their trends. Most factors affecting propagation are related to a 27-day solar cycle, the average time required for one complete rotation of the sun. Areas of solar activity, such as sunspots, may last several months, and face the earth more than once. These cycles become apparent if the values for solar flux and the A index are plotted over time. If you track these numbers on a daily basis, the current band conditions should come as no surprise. With some practice, you should be able to predict what conditions will exist next week or next month.

All this valuable data is of even more importance when applied to the propagation charts found in the various amateur magazines. The real-time information available through WWV or

WWV/WWVH Propagation Forecasts

All amateurs have a useful tool at their disposal for determining the current conditions of the ionosphere. These are the broadcasts from the National Institute of Standards and Technology stations WWV and WWVH. In addition to providing precise time signals and tones, these stations broadcast information on the current state of the ionosphere and 24 hour forecasts of expected propagation conditions. These "geoalerts" are broadcast at 18 minutes past the hour on WWV and 45 minutes past the hour on WWVH. Five different pieces of information are included with each transmission: solar flux as measured daily at 2000 UTC, Boulder A index, Boulder K index, solar activity and geomagnetic field conditions. Each has a bearing on the present and future conditions of the amateur HF bands.

Solar flux, as discussed previously, is one measure of the sun's activity. The minimum solar flux levels range in the lower 60s, corresponding to a quiet sun. These low readings normally occur only during the few years of the sunspot minimum. A flux reading of 63 or 64 usually coincides with a sunspot count of zero. Most of the time flux readings are above this level, and during the years of peak solar activity may exceed 250.

WWVH can help overcome the primary deficiency in any of these charts: The average solar flux figure on which any of these charts is based may not be a good representation of the large range of solar activity that can take place within a 28-31 day period. Additionally, the charts must be prepared months in advance, and the estimated values used for the charts can be quite different than the actual level of solar activity for the forecast period.

Know When and Where to Look on Each Band

Each HF band has something to offer the QRPer. Unfortunately, low-power operation to many means using only the 10- and 40-

meter bands. While 28 MHz may be the QRP dream band, it is not without its faults. Limiting your operating to one band or mode limits your activity in this most interesting of all amateur pursuits.

Imagine the DX operator who only works 40 meters, or the contester who limits himself to 160-meter single-band entries. While they both have fun, they are missing a larger segment of their chosen avocations. The QRPer, whether DXer, contester, or casual operator, cannot afford to lose out on opportunities simply because they occur on other bands.

Let's take a look at the popular HF bands from the viewpoint of a QRP operator. Each has its own individual propagation characteristics, modes of operation, and particular places where low-power operators hang out. Complete expertise requires you to spend some time operating on each and every one.

Familiarity with each of the bands is a long-term learning process. A single contest or one week's vacation spent chasing DX will not make you an expert. The ability to consistently wring contacts out of any band only comes with several years of intensive operating. Some QRPers have literally spent their lives learning the nuances of each band. For you, this chapter should speed up that process.

160 Meters (1.8 MHz)

Most hams view the "top band" with an equal mixture of dismay and despair. Everybody knows this is amplifier country, and you're going to spend your evenings getting an ear full of static.

Even limited operation on 160 meters during a DX contest provides the opportunity for a few extra multipliers and a couple of DX contacts. The multipliers are worth much more than the individual contacts, and can make the difference of whether you win or lose.

At one time, QRP signals were never heard on 160 meters. However, more and more rigs now include 160. With more people on this band, the QRPer finds increased chances of working DX and making contest contacts.

Not all countries have full use of the band. There are two DX windows, at 1825-1830 kHz and 1850-1855 kHz. Split-frequency operation is common. The DX station transmits inside the window, while stations in North America transmit on a frequency announced by the DX station (outside the window). Another DX window, used by Japanese stations, is located at 1907.5-1912.5 kHz. JAs usually listen on the low end of the band.

Activity generally follows the ARRL band plan: CW only in the lower 30 kHz and both CW and SSB from 1830-2000 kHz. Voluntary compliance to this band plan is generally followed by all stations using the band. If you're serious about chasing DX on 160 meters with QRP, you must be a CW operator. CW maximizes the effectiveness of your low power.

160 meters is strictly a nighttime band. The best operating conditions are during the winter months when atmospheric noise is lower and the MUF drops to lower frequencies than in summer. During years of little sunspot activity, gray-line propagation offers the best DX opportunities. From the US, DX conditions peak toward Europe around midnight, and swing around to the South Pacific just before sunrise. During years of higher solar activity, look for nighttime openings up to 1500 miles and occasional DX opportunities between local midnight and sunrise.

QRP operation is successful on 160 meters because there are many excellent DX stations. Someday it will be possible for a QRPer to work DXCC on this band, thanks to increased activity by these well-equipped high-power stations.

80 Meters (3.5 MHz)

To successfully chase CW DX on 80 meters, you need an Extra Class ticket in your pocket. The majority of DX stations found on 80 meters only operate in those portions of the band available to this license class. As a trade-off, that greatly simplifies your search for DX. Most DX stations operating CW are in the lower 25 kHz. Actually, 90% will be found from 3500-3515 kHz.

BEAT EVERYBODY UP TO WORK 80 METER DX

Split-frequency operation is common only during major DXpeditions. As with 160 meters, CW is the mode of choice for the QRPer.

Unlike CW, SSB DX operation is normally split frequency. DX stations may transmit in many portions of the band, while listening from 3790-3800 kHz. For the stateside QRPer, an Advanced or Extra Class license is the ticket to work them. This is the most challenging band for QRP SSB.

DXing 80 meters is unlike most other bands in that opportunities come and go with very specific openings. It is primarily a nighttime band. During the winter months there is little noise, and worldwide QSOs are possible. Cross-country contacts are a common occurrence.

During the years of maximum solar activity, DX possibilities from the US peak toward Europe and the east from an hour or so after sunset until midnight. Between midnight and dawn, the band opening swings around toward the west. It peaks to Asia and Oceania at sunrise. During the years of lower solar activity, 80 meters should be the best DX band at night. DX should be available from all parts of the world located in the nighttime hemisphere, with conditions peaking around midnight and again at sunrise.

The best opportunities for the QRPer are those short periods of time that your station lies within the terminator separating the daylight and darkness hemispheres. Of the two daily times, sunset and sunrise, the best opportunities come early in the morning. Chasing DX on 80 meters means setting your alarm clock and operating before the rest of the family is out of bed.

The QRP contester without 80-meter capability is just begging to be an also-ran. Eighty meters provides real opportunities to pick up multipliers when the level of competition is greatly reduced. This is particularly true on the US West Coast. In the early morning hours, the band is gone for the East Coast, which drastically reduces the QRM and competition for Pacific and Asian stations.

40 Meters (7 MHz)

There are a number of advantages to operating on 40 meters. It is a relatively reliable band, the noise level is down from 80 and 160 meters, antennas are smaller, and more DX opportunities abound.

Like 80 meters, chasing DX on 40-meter CW is a relatively simple affair. Most of the DX is found in the lowest 30 kHz of the band. More often than on 80 meters, DX operators will provide opportunities above 7025 kHz. SSB DX is, most often, a split-frequency affair. Typically, the DX station will transmit between 7050 kHz and 7100 kHz while listening above 7150 kHz. Opportunities for the SSB QRPer are enhanced because of normally better propagation conditions here than on 80 meters. Noise levels on 40 meters tend to be seasonal. The best band conditions are during the winter months.

The daylight hours are probably the best for chasing a 40-meter endorsement for the Worked All States (WAS) award. In the early evening, the skip distance increases, and the band opens up to distances of several thousand miles.

Worldwide communications are almost routinely possible during the evening and early morning hours. However, the setting sun brings a new type of interference: megawatt international broadcast stations operating from 7100 kHz to 7300 kHz. Propagation opportunities to Europe can be quickly checked by listening for these stations. They tend to be spaced every 10 kHz up the band from 7100 kHz. Below 7100 kHz, amateurs own the band worldwide and there's little interference from other radio services.

During the years of low solar flux, 40 meters really shines as a DX band. Many DX operators once found on 20, 15 and 10 meters migrate to 40 meters because of reduced high-band propagation. When solar-flux levels remain consistently below 100, watch 40 meters blossom with signals from all over the world. DX openings from the US to Europe and Africa begin in the late afternoon and swing to the south at sunset. The band remains open for several more hours until the MUF drops below 7 MHz. Openings to the South Pacific and Asia begin after midnight and peak at local sunrise.

During years of high solar activity, worldwide DX is possible to most parts of the world during the hours of darkness. The band opens in the late afternoon toward the east, with signals peaking to the south around midnight. Signals from Oceania and Asia are heard in the US in the early morning hours (especially if the sun is just setting at the other end), and peak just before dawn.

Even during the years of high solar activity, 40 meters provides consistent nighttime openings to many parts of the world. Lower noise levels, and the fact that many DX contesters have 40-meter beams, means you can accumulate a surprising number of contacts and multipliers.

30 Meters (10.1 MHz)

Thirty meters is a transition band; it shares characteristics with both 40 and 20 meters. It is also a shared band, where amateurs have a secondary allocation to other radio services. The QRP operator has one real advantage when operating this band: The maximum power that any US station may run is 200 watts output.

Since the band is so small (50 kHz wide), DX opportunities can be found almost anywhere. DX QSOs on 30 meters are just that. They are seldom the signal report and goodbye type of exchange commonly found on other bands. Thirty meters provides the greatest opportunities per kilohertz to chew the rag with DX stations.

Propagation on 10 MHz is a mixed bag. During years of low solar activity, the band opens earlier and closes later than 40 meters. From the US, the band opens toward Europe during the middle of the afternoon, swings around to the south at midnight, and peaks into the South Pacific during the early morning hours. During years of peak solar activity, it may be open to some area of the world at any hour of the day or night. At this time, DX opportunities peak toward Europe in the early afternoon and to Oceania in the midmorning.

Thirty meters isn't used during contests, and contacts made on 30 meters aren't usable for DXCC. Even though the contacts are just for fun, many DXpeditions show up on 30, usually near the end of their stay. The day after a major contest, like the CQ World Wide CW and SSB tests, the same stations you fought to work on 20 or 15 meters sometimes show up on 30, and are usually easier to work.

20 Meters (14 MHz)

If you had only one band to use for QRP DXing, it should be 20 meters. This is *the* DX band. Year in and year out, 20 meters provides the most DX opportunities. Eventually, every DX station in the world shows up on 20 meters.

When solar activity is high, 20 meters is often open 24 hours a day, providing a nonstop DX banquet. Peak conditions continue for several hours after sunrise, and again during the late afternoon and early evening hours.

Even during minimum solar activity, 20 meters is open during local daylight hours to most parts of the world. Peak conditions to the east occur for several hours after sunrise. The band peaks again, to the south and west, in the late afternoon.

Twenty meters has specific openings to selected areas that are almost as predictable as those on 80 and 40 meters. These band openings are more noticeable during years of lower sunspot activity.

Don't forget to check for long-path openings at local sunrise and sunset. Many long paths are mostly over water, rather than over land. Water reflects radio waves better than soil, so long-path signals are often quite strong.

Thanks to the Northern California DX Foundation (NCDXF), you can check 20-meter propagation to any part of the world any time, day or night. NCDXF coordinates an international beacon network on 14.100 MHz. Packet-station interference is sometimes a problem on this frequency, however.

Each beacon transmits a signal that is sequentially reduced from 100 watts to 0.1 watt. Transmissions last about one minute, and all beacons operate on a ten minute repeating cycle. Beacon antennas are either quarter-wave verticals or phased quad loops, giving an omnidirectional radiation pattern. The astute operator, by listening to these transmissions, can determine current DX possibilities to almost any place in the world.

17 Meters (18 MHz)

More and more DXpeditions are showing up on 17 meters, which is definitely helpful to QRP operators. Many hams don't have equipment to operate on this band, so the pileups are usually smaller than on 20 or 15 meters. If you're serious about DXing, you should pay careful attention to 17 meters.

Propagation conditions on 17 meters are similar to those on 20 and 15 meters, but 17 is often open when 15 is not. If the MUF is hovering around 19 or 20 MHz, you'll find better conditions on 17 than on 20 meters, and 15 meters won't be open at all. Like 30 and 12, 17 meters is not used for contests. If you're not looking for a winning score in a popular contest, you can often turn up some good DX on 17 meters, because many DX stations go to this band to avoid the contesters! Propagation beacons are found at 18.1 MHz.

15 Meters (21 MHz)

When solar activity is high, 15 meters offers more DX opportunities than 20 meters. Like 20, 15 meters may even be open all day and night. Absorption is lower than on 20 meters, and antennas can be smaller.

During much of the solar cycle, however, 15 meters opens and closes as the terminators cross your location. During peak solar activity, 15 meters opens from the US toward Europe at sunrise, peaks toward the south and southwest during the early afternoon, and to the Far East during the late afternoon and early evening.

During the years of low solar activity, openings from the northern hemisphere are primarily to the south from local noon to sunset. Openings on 15 meters tend to be more geographically specific than on 20 meters. While 20 provides continental openings, openings on 15 meters may be only to specific countries.

The best long-path openings occur after a minimum of one hour of darkness. From the US, the best times seem to be around 2100 local, with the beam pointed to the south or southeast. Another opening occurs to the southwest, just after local sunrise. Around the fall and spring equinoxes (September 21 and March 21), the northern and southern hemispheres receive nearly equal solar radiation. Fantastic openings between the hemispheres often occur at these times of the year. Finally, 15 meters is worth checking around midnight for an occasional opening in a southwesterly direction to your antipode (that region of the earth farthest from you).

Because 15 meters is wider than either 20 or 40 meters, there is less concentration, by mode, of DX stations. CW DX operators are normally spread out over the lower 75 kHz of the band. The more common DX countries are found higher in the CW subband, while the really rare stuff operates in the lower 25 kHz of the band—*usually*. It pays to check the whole band. DX can also be found between 21.1 and 21.2 MHz. A number of Novices and Technicians have worked DXCC on their 15-meter CW subband. One place where everybody should look for rare DX is the upper 25 kHz of the Novice/Technician subband. DX stations may be found engaging in ragchew QSOs, far from the crowds on 21.025 MHz.

You'll find more opportunities for QRP SSB DX on 15 than on 20 meters. There is a lower level of QRM, and normally less competition from high-power stations. DX is mostly found below 21.4 MHz. A favorite frequency for DXpeditions is 21.295 MHz.

A Propagation Primer

12 Meters (24.9 MHz)

Like 17 meters, the 12-meter band is a good place to work DX with QRP. While its population is rising steadily, you have a much better chance of working a DX station that shows up here than you do on 10 or 15 meters. Propagation conditions resemble those on 15 and 10 meters, with which it shares the same relationship that 17 meters has with 15 and 20. Twelve meters is not used in contests.

10 Meters (28 MHz)

When 10 meters is open, it seems that almost any level of radiated RF energy produces contacts. QRM is almost nonexistent, due to the large size of the band and its unique propagation characteristics.

When the band is dead, it's really dead. At these times, 10 kW to phased rhombics won't produce DX contacts. During the years of minimum solar activity, you can tune the entire 1700 kHz of this band and almost never hear anything except for local stations.

But while 10 meters does come and go with the 11-year sunspot cycle, there is activity during the lean years of low flux. Sporadic-E propagation shows little regard for recorded levels of solar activity.

There are numerous 10-meter beacons all over the world. Most of them run only a few watts and, given the normally reciprocal conditions of the band, if you hear them, you can work into that part of the world.

Ten meters is mostly a daylight band. Propagation openings follow the sun. During peak solar activity, conditions from the US peak toward Europe and Africa in the late morning hours. Early afternoon brings openings to the south, and during the late afternoon and early evening, the band opens to the South Pacific and Asia.

During those lean years of little solar activity, DX possibilities on 10 meters are very poor. There are occasional openings to South America and other tropical areas during the daylight hours from late fall to early spring. During the spring and early summer months, sporadic-E activity may provide a number of short-skip openings between 300 and 1300 miles.

The long-haul circuits found on 10 meters during high solar

activity are seldom open for very long. For example, the path between the Pacific Northwest and Europe is only open for several hours a day. Polar paths seem to live and die during specific time periods. Much of the time, it is relatively easy to hear this window of opportunity as it sweeps across the North American continent.

Trans-equatorial paths, on the other hand, tend to open for a much greater part of the day. Openings from the US to South America and Oceania may begin in the early morning and last well beyond local sunset.

Surprisingly, during the years of maximum solar activity, there are occasional long-path openings on 10 meters. While they are rare, when they do occur it's like talking to a DX station on the telephone. From the US, check to the south and southeast just after local sunrise, and the south and southwest several hours after local sunset. These paths cover Africa, the Middle East, the Far East and Europe, depending on your location.

It's also a good idea to look to the southeast (from the West Coast) or southwest (from the East Coast) just after local sunrise. (Stations in the Midwest should adjust beam headings accordingly.) There may be a short-term, long-path opening into the Indian Ocean area. However, this path quickly gives way to the normal short-path opening.

Ten meters provides more opportunities for the sideband QRPer than for those who operate only on CW. More DX shows up on SSB than on CW. Most DX sideband activity is concentrated between 28.45 MHz and 28.55 MHz. There is very little split-frequency operation on 10 meters. Its use is limited primarily to DXpeditions.

CW DX may be found anywhere in the lower 100 kHz of the band. There are no exclusive Amateur Extra class subbands on 10 meters. As with the other HF bands, the rarest DX is usually found in the lower 25 kHz, but they're fair game for everyone.

Notes
[1]MINIPROP is available from W6EL Software, 11058 Queensland St, Los Angeles, CA 90034-3029.
[2]XANTEK Inc, PO Box 834, Madison Square Garden Station, New York, NY 10159.

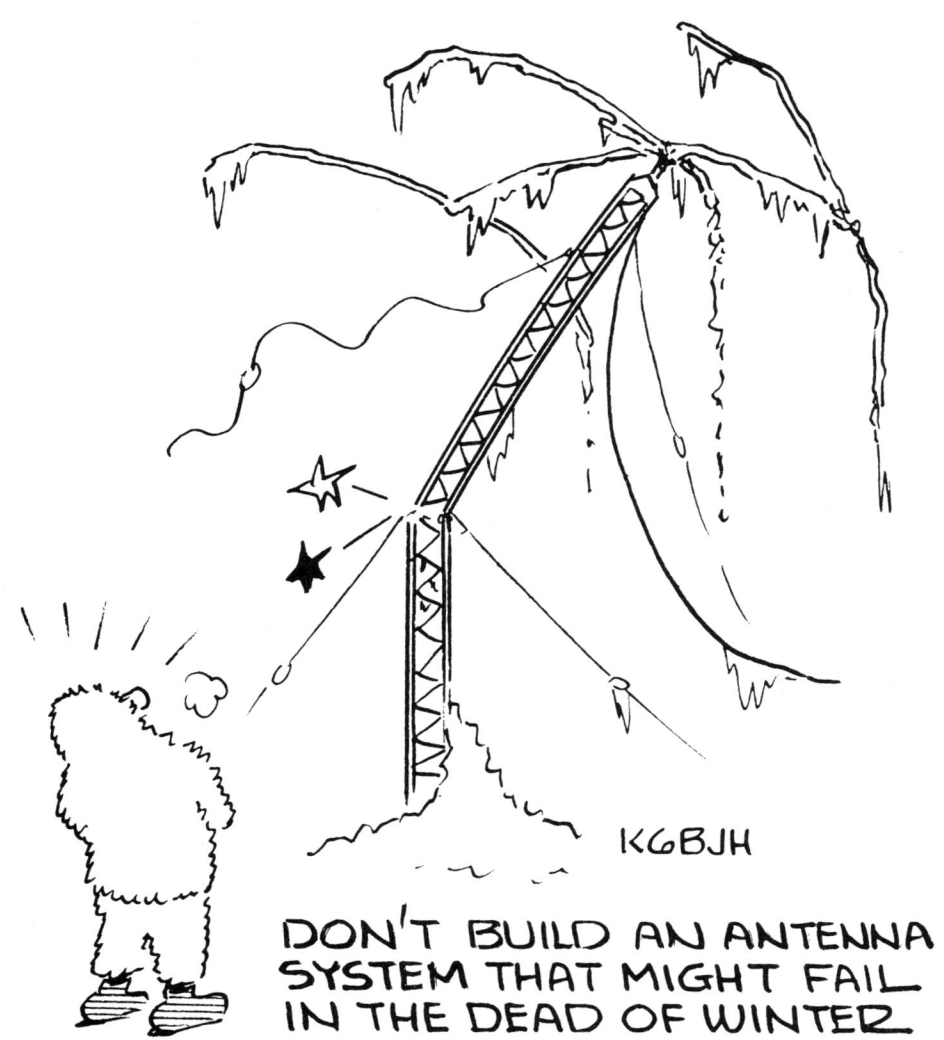

CHAPTER 5
Maximize Your Signal

Perhaps the single most important piece of equipment at a QRP station is the antenna. Next to the ground system, it is often the most neglected piece of equipment. Far too often, operators handicap their efforts by equating low power with poor antennas.

How often have you heard or read about QRP exploits where the operator was running with a random-wire antenna? It seems that many who write about QRP preface their articles with a story of how they worked some DX while their rig was attached to a metal door frame, set of bed springs, aluminum rain gutters, or some such mediocre antenna. It seems that we most often read of this in QRP construction articles.

Nothing against these articles or their authors, but they've got their priorities all screwed up. This book is primarily for QRPers interested in boosting their confirmed DX country totals and contest scores or being able to ragchew on a regular basis with stations outside their county. As such, our primary goal is multiple, rather than single, entries in the log book. It takes more than a piece of wire tossed out a window to build a fat log book.

The cardinal rule for QRP operating is: maximize your antenna system. Given the low power levels we use, you will make many more contacts if your antenna consists of stacked 8-element monobanders on a 120-foot tower than if you're pumping RF to a 50-foot random wire.

Injecting a bit of reality into this situation, let's modify our cardinal rule to read: Maximize your antenna system within the limitations of your pocketbook. While it's nice to dream and drool about stacked arrays on tall towers, most of us have additional financial commitments that take precedence over ham radio. I've still got four of those commitments at home. It's a shame that we can't make effective antennas from worn-out shoes or cast-off clothing.

Another factor that enters into the maximum antenna equation is where we live. A few of us are blessed with rural acreage, many live on city lots, while the balance dwell in apartments and condominiums. In addition to the size of our property, many homes are located in areas with restrictive covenants. You may have the misfortune to live in a housing development that allows no visible antenna structures or limits their height to something less than a quarter-wave on 10 meters.

Even those of us blessed with some available acreage have our share of problems. When Cathy and I purchased our current residence, I had three priorities: there had to be more than one acre, it had to be on the top of a hill, and there had to be several tall trees, at least 60 feet or better, within a hundred feet of the house. Well, I got two out of three. We have three acres and live on top of a hill. However, the nearest trees are some 200 feet from the house and consist of scrub alder about 30 feet high. My neighbors, of course, are not hams and they all have 150-foot fir trees close to their homes. Why is it that those who couldn't care less about 80-meter antennas always have the support structures?

So we further modify our axiom: Maximize your antenna system within the financial limitations of your pocketbook and the physical restrictions of your real estate.

WHAT KIND OF ANTENNA?

Rotatable, directional antennas are obvious choices wherever possible. If you can install Yagi antennas for 40 through 10 meters, so much the better. Multi-band beam antennas are the most popular HF antenna for DXing and contesting because they combine several bands in one low-profile antenna.

Separate Yagis for each band are a better choice if you can afford them (and your tower can support them). Single-band Yagis are free of the spacing and element-length compromises necessary in a multi-band design.

Vertical antennas for the bands above 14 MHz are not a good choice for serious QRP contesting and DXing. Because they aren't directional, verticals scatter your signal to the four winds, instead of pointing it at the stations you're trying to work. If space or economic reasons limit you to a vertical, install it as high as possible. It may work when mounted on the ground, but your signal will be partly absorbed by nearby buildings and trees. Ground-mounted verticals also tend to pick up more locally generated noise.

On the 160, 80 and 40-meter bands, a vertical may be the best choice for long-distance work. Depending on where you live, you may be able to work DX with QRP on 160 and 80 meters. It can be done from either coast during the winter months when static is lowest. On 40 meters, working DX with 5 watts into a vertical is much easier. A 40-meter vertical need only be about 33 feet high (quarter wavelength), so you may be able to mount it on the roof of your house. An elevated mount means the radials are also elevated, which reduces ground loss and improves performance.

Wire Antennas

A stack of gleaming beam antennas sure helps in the pileups, but not everyone can have them. Don't despair! If you can get some wire up in the air, you can still be competitive.

The simplest effective wire antenna is the dipole. If you feed a dipole with open-wire line and tune it with a Transmatch, you can use it on several bands. Like any antenna, dipoles work better when they're high and clear of obstructions, like buildings, trees and power lines. Do you have two trees over 100-feet high and more than 135 feet apart? If so, you'll be able to work plenty of DX on 80 meters. A lower 80-meter dipole is still usable, but the higher radiation angle makes working DX harder. For 40 meters, a dipole 50 feet or more in the air is very effective. When fed with open-wire line and tuned, it works very well up to 10 meters.

The inverted V is a variation of the dipole, where the ends slope toward the ground. Inverted Vs are harder to use on higher frequencies, and they're more directional than flat dipoles. That doesn't mean they don't work at all. If you only have one tall support, try one.

Loops are also effective DX antennas. A full-wave loop needs only one support, and can usually be fed with coaxial cable if you're only interested in one band. Using a loop on several bands is possible if you feed it with open-wire line and use a Transmatch. If possible, install a single-pole single-throw (SPST) relay at the corner opposite the feed point. Experiment with the loop open and closed, as its directional pattern and elevation angle will change. If you're only running QRP, you don't need a large relay. On some bands, the voltage across the open relay contacts is high; with over 100-watts output you need a pretty large contact spacing.

Multiband cubical quad antennas have one driven element for each band. On some quads, all driven elements are tied together and fed with one feed line. To maintain directivity, you can hang more than one loop from the same support and use one feed line. It's possible to make a fixed quad by hoisting a section of TV masting into the tree. The mast becomes the boom, and the elements are supported from the top. If the boom is long enough, you can even make a two-direction quad with a common reflector and two, switched driven elements. When you pick the directions, make sure they're pointing at places on the globe where there's ham activity!

Quads seem to work pretty well when close to the ground. It's important to try to clear obstructions, though. A quad beaming at a building won't work as well as one in the clear.

In addition to the quad, there are other good directional wire antennas. If trees are plentiful, try making wire Yagis. Yagis require some fiddling to tune properly, so be prepared to haul them up and down. Another good single-band wire antenna is the Sterba array. Sterbas require supports about 3/4-wavelength high. Again, higher is better. The length of a Sterba ranges from about 1 wavelength and up. You make a Sterba longer by adding more elements to the array. Doubling the number of elements approximately doubles the gain. Because you can't turn this array, don't make it *too* long. As the gain increases, the directional pattern becomes narrower. Instead of having a good signal into Africa, you may wind up with an excellent signal into only the Ivory Coast; the rest of Africa will be off the sides of your antenna's pattern. Sterba arrays about 2 wavelengths long are just about right.

Other useful, single-band wire arrays include the Bruce and the bobtail curtain. The books listed at the end of this chapter will give you many years worth of ideas to try.

Antennas for Field Day

Field Day antennas are a special case. You want antennas that work well, yet are easy to install and take down. Field Day offers no multiplier credits for working different states or sections. Instead, your goal should be to work as many stations as possible. If you live on or near either coast, directional antennas for the high bands, pointed toward populated areas 1000 to 2500 miles away, will provide contacts on those bands. On 80 and 40 meters you'll be satisfied to work stations less than 1000 miles away, meaning dipoles at less than 50 feet will work fine.

If you live in the central states, you need high-band antennas that cover both coasts, such as loops or other bi-directional gain antennas. Rotary beam antennas on masts are nice if you have a large enough crew to spare someone to turn them from time to time. Again, dipoles at about 50 feet will suffice for the low bands.

While it is tempting to carry a multi-band vertical out to the Field Day site, try to come up with a more directional antenna for the high bands. The interference on Field Day is legendary, and many Field Day operators are unskilled at contest-style operation. This means you need the loudest possible signal to get through to them. On the low bands, the vertical is more suited for long-distance operation, and will produce disappointing results on Field Day. Stick to dipoles or inverted Vs for the low bands.

BUILD IT TO LAST

An antenna system that fails in the dead of winter, when propagation is at its best, is useless. Reliability is often overlooked when we have a limited amount of money to spend and are shooting for a maximum gain system.

Going for greatest gain at the expense of reliability seems to most often befall contesters. I can personally attest to this fact. A year after being bitten by the contest bug, I scrounged some limited funds together to erect the biggest array I could afford.

I had the good fortune to find a 50-foot tower at a price of $1.00 per foot. Of course, it came out of a scrap metal yard, but no matter. Lots of elbow grease and some paint made it look almost as good as new. I opted to top it with a big 5-element triband (20, 15 and 10 meter) Yagi. This took all my money, but I figured the rotor I had used to turn my older 3-element antenna would suffice.

We live in a relatively mild climate and the whole system worked like a champ. I had all sorts of fun that winter working contests with 5 watts. Along about January, I noticed that the rotor was making more noise than usual when turning the beam. But it seemed to work okay, and the weather wasn't real conducive to climbing up the tower to check things out. Unfortunately, the rotor failed two days before the ARRL DX contest. Even that might have not been so bad except that the beam was pointed at 140 degrees. You don't know frustration until you've tried to work Europe and Japan with your antenna pointed in the

direction of the South Atlantic.

Due to bad weather, inertia, and financial commitments, it was another year before I could replace the rotor. The old one was beyond repair. When I opened its case, broken parts and ball bearings cascaded across the floor. It looked as though someone had taken a sledgehammer to its internal mechanism.

FEED LINES

The feed line that you use between the rig and antenna is of critical importance. If your antenna is fed with coaxial cable, use the best you can afford. Don't use a small cable such as RG-58. Its price is attractive, it is inconspicuous, and it can easily handle 5 watts. The problem with physically small cables is their high attenuation.

Coaxial Cable

Under normal conditions, the loss of RF energy in a coaxial cable is not noticed when you're running 100 or 1500 watts. There is no perceived difference between 1500 and 1475 watts at the

HARDLINE IS A LITTLE DIFFICULT TO WORK WITH

antenna. However, when you're only inserting 5 watts into a cable, the loss of even a single watt is noticeable. Coaxial-cable attenuation increases with frequency. While there might be no perceived loss on 80 meters, you could halve your signal on 10 meters.

The best coaxial cable is Hardline. Hardline is a semi-rigid coaxial cable with a solid (rather than flexible braided) outer conductor. Because it is rigid, a solid dielectric is unnecessary. Hardline cables often use insulating spacers to separate the inner and outer conductors. The result is much lower loss than that of flexible cable. Hardline diameters range from 1/2 inch to over 1 inch.

Hardline is not without its disadvantages. The first of these is cost. Hardline costs 4 or 5 times as much as flexible coaxial cable, though surplus cable-TV Hardline is often quite cheap. Hardline connectors though, are also expensive, and are not available as often on the surplus market. Being semi-rigid, Hardline is more difficult to work with than flexible coaxial cable.

If Hardline is out, buy the best flexible coax, such as RG-213. Cheaper coaxial cables have less shielding (the braid may cover only 50 percent of the dielectric, compared with 95 percent for RG-213) and thus more loss by radiation. Cheaper coaxial cables usually have outer jackets that deteriorate quickly when exposed to sun, wind and precipitation. The best cable is not much more expensive than the garden varieties. When you calculate the cost difference over the life of the cable, the good stuff is a bargain. What's more, the better cable will soak up less of your signal.

Type 9913 cable is similar in size to RG-213, but features a copper foil covering over the braided copper, providing a 100-percent shield. Instead of solid center insulation, it uses a spirally wrapped dielectric. The result is a lower attenuation than other cables of the same diameter. Type 9913 is often referred to as coaxial Hardline. It is the cable of choice for VHF-UHF operation. It is, however, more fragile than RG-213 and will not withstand repeated twisting.

Now that you've decided to buy the best coaxial cable, don't try to save money by using cheap connectors. Very often, the low-cost connectors are built of inferior materials and are incapable of surviving very long outdoors. Good-quality silver-plated connectors with Teflon insulation are not expensive if you shop around, and they are worth the extra money.

Open-Wire Line

Open-wire line offers very low attenuation, even compared to the best coaxial lines. If you have an antenna tuner, a single 80-meter dipole fed with open-wire line can be used on all higher-frequency bands. Even if you have a beam for the higher bands, the dipole is useful on 80, 40 and 30 meters. Open-wire line can't be run through a wall as easily as coaxial cable, and you have to keep it clear of towers, gutters and aluminum siding.

ANTENNA INSTALLATION TIPS

Don't forget to weatherproof every connection that is exposed to the outside elements. Coax-Seal is popular, but it's hard to get it off if you want to disconnect the cable. Double-sided silicone tape is sometimes available at hamfests. While more expensive than Coax-Seal, it's easier to remove.

When you're installing a tower, follow all of the manufacturer's recommendations. Don't skimp on any part of the construction job. Every tower is rated at the number of square feet of antenna it will support at a stated wind velocity. Play it safe by installing antennas that are smaller than the maximum load rating of the tower.

References

G. Hall, K1TD, *The ARRL Antenna Book*, 16th ed., 1991.
J. Lawson, W2PV, *Yagi Antenna Design*, Newington: ARRL, 1986.
L. Moxon, G6XN, *hf antennas for all locations,* London: RSGB, 1982 (available from ARRL).
J. Heys, G3BDQ, *Practical Wire Antennas*, London: RSGB, 1989 (available from ARRL).

Fadeout

I hope you've cranked down your power and are beginning to discover some of the many things that can be accomplished with a limited amount of RF energy. By now you've experienced some of the thrills of low-power operation: emerging bloody, but victorious, from the screaming chaos of a pile-up; capturing a rare DX station by the stealthy tuning of a band; the successful reconnaissance of foreign territories during a long-path opening.

Every time you turn on the rig and get on the air, it should be a learning experience. Remember what worked and what didn't in a particular situation. Savor this knowledge and apply it the next time you operate. You're involved in an aspect of Amateur Radio that places a heavy emphasis on experience. If you persevere, you will develop an expertise that is all out of proportion to your power level.

I would again point out to those of you who, for whatever reason, do not operate contests that you are missing a unique opportunity to work DX. Any contest represents weeks of normal operating compressed into a time frame measured in hours. In 60 minutes of any contest, you can work DX that could otherwise take a week or month of patient searching through the bands. Many countries that are normally considered rare appear with regularity during these contests and are relatively easy to work. Should they show up at any other time, they immediately generate huge, screaming pile-ups that will send the average QRPer to his nearest drugstore for a bottle of Rolaids.

Finally, keep a sense of perspective. Even though I operate QRP with a combat mind-set, I realize that it isn't a matter of life-or-death. There are times when you must surrender gracefully to overwhelming competition or poor band conditions. However, there's always tomorrow and, while you may lose a skirmish, you will be the ultimate winner in this global war of watts.

Appendix A

QRP Clubs and Nets

QRP Clubs

QRP Amateur Radio Club International
Mike Kilgore, KG5F
2046 Ash Hill Road
Carrollton, TX 75007
New membership: $12.00
Newsletter: *QRP Quarterly*

Michigan QRP Club
c/o Membership Chairperson
5346 West Frances Road
Clio, MI 48420
Annual Membership: $7.00
Newsletter: *The Five Watter*

G-QRP
US Agent: Luke Dodds, W5HKA
2852 Oak Forest
Grapevine, TX 76051
Annual Membership (US): $12.00
Newsletter: *Sprat*

QRP Nets

Net	Frequency (kHz)	Day (UTC)	Time (UTC)
TCN	14060	Sunday	2300
SEB	7030	Wednesday	0001
GSN	3560	Thursday	0200
GLN	3560	Thursday	0200
NEN	7040	Saturday	1300
WSN	7040	Saturday	1600
MI-QRP	3535	Wednesday	0100
VE-QRP	14060	Sunday	1900
NWQRP	7040	Tuesday	0400

Appendix B
Contests With QRP Categories

All of the contests listed below allow for QRP competition. Many of them also allow higher levels of power, but this is a different operating category within the individual contest. In any of these contests, you are only competing against other operators running five watts or less. The guidelines and rules listed for each of these contests are subject to change. I would encourage you to write the individual contest sponsor for the latest information. This listing is not all-inclusive, but a representative sample of the more popular contests.

CQ WORLD-WIDE WPX CONTEST
Sponsor
CQ Magazine
Dates
SSB: last full weekend in March
CW: last full weekend in May
Contest Period
0000 UTC Saturday to 2400 UTC Sunday. Only 30 hours of the 48 hour contest period permitted for Single Operator stations.
Objective
Everybody works everybody
Bands
160-10 meters. No WARC bands
Type of Competition (QRP)
Single Operator, All Band and Single Operator, Single Band
Exchange
RS(T) report plus a progressive 3-digit serial number starting at 001.
Points
Contacts between station on different continents are worth three points on 28, 21, and 14 MHz and six points on 7, 3.5, and 1.8 MHz. Contacts between stations on the same continent are worth one point on 28, 21, and 14 MHz and two points on 7, 3.5, and 1.8 MHz. For North American stations only: contacts between station within the North American boundary count as two points on 28, 21, and 14 MHz and four points on 7, 3.5, and 1.8 MHz. Contacts between stations in the same country are permitted for multiplier credit only.
Multipliers
The number of different prefixes worked on all bands. A prefix is counted only once, regardless of band.
Scoring
Total QSO points from all bands

3

multiplied by the number of different prefixes worked
Awards
Certificates to the highest scoring station in each country and in each call area of the United States, Canada, Australia, and Asiatic USSR. There are additional trophies and plaques given by individual donors.
Information and Logs
CQ Magazine
76 North Broadway
Hicksville, NY 11801

CQ WORLD-WIDE DX CONTEST
Sponsor
CQ Magazine
Dates
SSB last full weekend in October
CW last full weekend in November
Contest Period
0000 UTC Saturday to 2400 UTC Sunday.
Objective
Everybody works everybody
Bands
160-10 meters. No WARC bands
Type of Competition (QRP)
Single Operator, All Band and Single Operator, Single Band
Exchange
RS(T) report plus *CQ* Zone number.
Points
Contacts between stations on different continents are worth three points. Contacts between stations on the same continent are worth one point. For North American stations only, contacts between stations within the North American boundary are worth two points. Contacts between stations in the same country are permitted for zone or country multiplier credit, but have zero point value.
Multipliers
A multiplier of one for each different zone contacted on each band. A multiplier of one for each different country contacted on each band. Stations are permitted to contact their own country and zone for multiplier credit.
Scoring
The total QSO points multiplied by the sum of your zone and country multiplier
Awards
Certificate for the top scoring station in each country and in each call area of the United States, Canada, European Russia, Asiatic USSR, Czechoslovakia, and Japan. To be eligible for an award, a Single Operator station must show a minimum of 12 hours of operation. Additional trophies and plaques by individual donors.
Information and Logs
CQ Magazine

ARRL INTERNATIONAL DX CONTEST
Sponsor
The American Radio Relay League
Dates
CW: Third Full Weekend in February
SSB: First Full Weekend in March
Contest Period
0000 UTC Saturday to 2400 UTC Sunday
Objective
W/VE works DX, DX works W/VE
Bands
160-10 meters, except 30, 17 and 12 meters
Type of Competition (QRP)
Single Operator, All Band only
Exchange
W/VE stations send RS(T) and state or province. DX stations send RS(T) and three-digit number indicating output power.
Points
W/VE: each DX QSO counts three points
DX: each W/VE QSO counts three points

Multipliers
W/VE stations: Sum of DXCC countries worked (except US and Canada) per band
DX stations: Sum of US states (except Hawaii and Alaska) and District of Columbia, VE1-8, VO, VY1 worked per band. Maximum of 59 per band.
Scoring
Total QSO points multiplied by the sum of multipliers per band
Awards
Certificates awarded to top Single Operator-All Band entry from each country and ARRL Section. Plaques to top W/VE and DX QRP entry for both phone and CW. Other special plaques and trophies by individual donors.
Information and Logs
ARRL
225 Main Street
Newington, CT 06111

ARRL SWEEPSTAKES
Sponsor
The American Radio Relay League
Dates
CW: First full weekend in November
SSB: Third full weekend in November
Contest Period
2100 UTC Saturday to 0300 UTC Monday
Objective
W/VE work W/VE only.
Bands
160-10 meters, except 30, 17 and 12 meters
Type of Competition (QRP)
Single Operator, All Band.
Exchange
A consecutive serial number, Precedence ("Q" for five-watts or less output), your call, check (last two digits of year first licensed), and your ARRL Section.

Points
Two points per QSO. Work each station only once.
Multipliers
Each ARRL Section and CRRL Section plus VE8/VY1. Maximum of 77 multipliers.
Scoring
Multiply QSO points by Number of Sections worked
Awards
Certificate for top QRP station in each ARRL and CRRL Section and Division
Information and Logs
ARRL

ARRL FIELD DAY
Sponsor
The American Radio Relay League
Dates Fourth full weekend in June.
Contest Period
1800 UTC Saturday to 2100 UTC Sunday. Operate 24 consecutive hours of the contest period.
Objective
W/VE work W/VE
Bands
All bands except 30 meters
Type of Competition (QRP)
Field Day entries are classified according to the maximum number of simultaneous transmitted signals followed by a class designator. See the rules in May *QST*.
Exchange
Field Day Operating Class and ARRL/CRRL Section
Points
One point for each SSB/FM contact and two points for each CW or Digital contact
Multipliers
None
Scoring
Sum of contact points on all bands times power multiplier. QRP with battery power source, multiply by five (5).

5

Bonus Points
100 points per transmitter for 100% emergency power
100 points for media publicity
100 points for locating in a public place
100 points for public-relations display
100 points for message to SM or SEC
10 points each for maximum of 10 messages relayed
100 points for satellite QSO
100 points for natural power (wind, solar, water)
100 points for copying W1AW Field Day message
100 points for at least one packet QSO (via digipeater okay)
Awards
None (Field Day isn't considered a contest)
Information and Logs
ARRL

AGCW-DL QRP WINTER CONTEST
Sponsor
Activity Group CW-DL
Dates
Third weekend in January
Contest Period
1500 UTC Saturday to 1500 UTC Sunday
Objective
Everybody works everybody
Bands
160-10 meters
Type of Competition (QRP)
CW Only: Class A (single operator, 2 watts output)
Class B (single operator, 5 watts output)
Class C (multioperator, 5 watts output).
Class C operates full 24 hours.
Class A and B must break for nine hours, which may be split into 2 equal segments
Exchange
RST, consecutive QSO number, and power input

Points
One point for QSO with own country
Two points for QSO with own continent
Three points for QSO with different continent
Multipliers
DXCC countries and call areas in JA, PY, VE, W, and ZS
Scoring
Band score equals QSO points times multipliers. Total score equals sum of band scores. Separate logs for each band.
Information and Logs
Sigfried Hari, DK9FN
Spessartstrabe 80
D-6453 Seligenstadt
Germany

IRSA WORLD RADIO CHAMPIONSHIP
Sponsor
International Radiosport Association
Dates
First weekend in October
Contest Period
SSB - 0000Z-2400Z Saturday
CW - 0000Z-2400Z Sunday
Note: These are separate contests
Objective
Everybody works everybody
Bands
160-10 meters, except 30, 17 and 12 meters
Type of Competition (QRP)
Single Operator, All Band and Single Operator, Single Band
Exchange
RS(T) and progressive serial number beginning with 001
Points
Two points for SSB QSO, three points for CW QSO
Multipliers
Each DXCC country and call areas in each of the following countries worked

on each band: USA, Canada, Australia, Brazil, Japan, European and Asiatic USSR
Scoring
Total points from all bands times the total multipliers from all bands
Awards
Certificates to winners in all categories in all participating DXCC countries and call areas. Additional plaques by individual donors.
Information and Logs
Walt McGugan, W3FG
P. O. Box 7
Odenton, MD 21113-0007

RSGB 21 MHz CW CONTEST
Sponsor
Radio Society of Great Britain
Dates
Third Sunday of October
Contest Period
0700-1900 UTC
Objective
Everybody works the British Isles
Bands
21 MHz only.
Type of Competition (QRP)
Single Operator.
Exchange
RST plus progressive serial number beginning with 001.
Points
Three points for each G, GB, GI, GM and GU station worked
Multipliers
Number of G prefixes worked, up to a maximum of 49
Scoring
Total QSO points multiplied by total prefixes worked
Awards
Certificates to three top winners and, at the discretion of the contest committee, to the leading station in each country
Information and Logs
RSGB Contest Committee
P. O. Box 73
Lichfield
Staffs WS13 6UJ
England

ARCI QRP FALL CW CONTEST
Sponsor
QRP Amateur Radio Club International.
Dates
Third weekend in October
Contest Period
1200 UTC Saturday to 2400 UTC Sunday. Operate 24 hours maximum.
Objective
Everybody works everybody
Bands
160-6 meters, except 30, 17 and 12 meters
Type of Competition (QRP)
Single Operator, All Band
Single Operator, Single Band.
Exchange
RST and state, province, or DXCC country. Members include their number, non-members their power output.
Points
Five points for member contact. Two points for others on same continent and four points for others on different continent.
Power Output Bonus
0 to 1 watt: Multiply final score by 10
1 to 5 watts: Multiply final score by 7
Over 5 watts: Check log
Solar or wind power for duration of contest: Multiply final score by 2
Battery power for duration of contest: Multiply final score by 1.5
Homebrew equipment bonus (per band): 2000 points for transmitter, 3000 points for receiver, 5000 points for transceiver

Multipliers
Each state, province, or DXCC country worked on each band.
Scoring
(Total QSO points from all bands) × (sum of states, provinces, countries) × (power output bonus) × (power type) + (homebrew bonus).
Awards
Certificates to the top ten scores overall
Top score on each band for single band entries
Top score in each state, province, and DXCC country.
Information and logs:
Red Reynolds, K5VOL
835 Surryse Road
Lake Zurich, IL 60047

QRP ARCI WINTER FIRESIDE CHAT
Sponsor QRP Amateur Radio Club International
Dates
Third Sunday in January
Contest Period
2000—2400 UTC.
Bands
160-6 meters, except 30, 17 and 12 meters
Type of Competition (QRP)
SSB Only: Single Operator, All Bands
Single Operator, Single Band.
Exchange
RS and state, province, or DXCC country. Members include their number, non-members their power output.
Points
Five points for member contact. Two points for others on same continent and four points for others on different continent. Power output bonus: 0 to 1 watt—× 10.
1 to 5 watts—× 7
Over 5 watts is check log. Solar or wind power for duration of contest —× 2. Battery power for duration of contest—× 1.5
Multipliers
Each state, province, or DXCC country worked on each band.
Scoring
(Total QSO points from all bands) × (sum of states, provinces, countries) × (power output bonus) × (power type).
Awards
Certificates to the top ten scores overall
Top score on each band for single band entries
Top score in each state, province, and DXCC country.
Information and logs:
Red Reynolds, K5VOL
835 Surryse Road
Lake Zurich, IL 60047

QRP ARCI HOLIDAY HOMEBREW SPRINT
Sponsor
QRP Amateur Radio Club International
Dates
Second Sunday in December
Contest Period
2000 - 2400 UTC.
Bands
160-6 meters, except 30, 17 and 12 meters
Type of Competition (QRP)
CW Only: Single Operator, All Bands
Single Operator, Single Band.
Exchange
RST and state, province, or DXCC country. Members include their number, non-members their power output. Add suffix of "HB" to RST if home brew gear. Add suffix of "C" if commercial gear.
Points
Five points for member contact. Two points for others on same continent and

four points for others on different continent. Power output bonus: 0 to 1 watt—× 10; 1 to 5 watts—× 7; over 5 watts is check log. Solar or wind power for duration of contest—× 2; Battery power for duration of contest—× 1.5; Homebrew equipment bonus (per band): 2000 points for transmitter, 3000 points for receiver, and 5000 points for transceiver.
Multipliers
Each state, province, or DXCC country worked on each band.
Scoring
(Total QSO points from all bands) X (sum of states, provinces, countries) X (power output bonus) X (power type) + (homebrew bonus).
Awards
Homebrew gear required for contest entry. Commercial equipment is check log. Certificates to the top ten scores overall Top score on each band for single band entries
Top score in each state, province, and DXCC country.
Information and logs:
Red Reynolds, K5VOL
835 Surryse Road
Lake Zurich, IL 60047

QRP ARCI HOOT OWL SPRINT
Sponsor
QRP Amateur Radio Club International.
Dates
Fourth Monday in May
Contest Period
2000 - 2400 LOCAL time.
Objective
Everybody works everybody.
Bands
160-6 meters, except 30, 17 and 12 meters
Type of Competition (QRP)
Single Operator, All Band
Single Operator, Single Band.

Exchange
RST and state, province, or DXCC country. Members include their number, non-members their power output.
Points
Five points for member contact. Two points for others on same continent and four points for others on different continent. Power output bonus: 0 to 1 watt—× 10; 1 to 5 watts—× 7; over 5 watts is check log. Solar or wind power for duration of contest —× 2; Battery power for duration of contest—× 1.5; Homebrew equipment bonus (per band): 2000 points for transmitter, 3000 points for receiver, and 5000 points for transceiver.
Multipliers
Each state, province, or DXCC country worked on each band.
Scoring
(Total QSO points from all bands) × (sum of states, provinces, countries) × (power output bonus) × (power type) + (homebrew bonus).
Awards
Certificates to the top ten scores overall Top score on each band for single band entries
Top score in each state, province, and DXCC country.
Information and logs:
Red Reynolds, K5VOL
835 Surryse Road
Lake Zurich, IL 60047

QRP ARCI SPRING QSO PARTY
Sponsor
QRP Amateur Radio Club International.
Dates
Second Weekend in April
Contest Period
1200 UTC Saturday to 2400 UTC Sunday.

Objective
Everybody works everybody.
Bands
160-6 meters, except 30, 17 and 12 meters
Type of Competition (QRP)
Single Operator, All Band
Single Operator, Single Band.
Exchange
RST and state, province, or DXCC country. Members include their number, non-members their power output.
Points
Five points for member contact. Two points for others on same continent and four points for others on different continent. Power output bonus: 0 to 1 watt—× 10; 1 to 5 watts—× 7; over 5 watts is check log. Solar or wind power for duration of contest—× 2; Battery power for duration of contest—× 1.5; Homebrew equipment bonus (per band): 2000 points for transmitter, 3000 points for receiver, and 5000 points for transceiver.
Multipliers
Each state, province, or DXCC country worked on each band.
Scoring
(Total QSO points from all bands) × (sum of states, provinces, countries) × (power output bonus) × (power type) + (homebrew bonus).
Awards
Certificates to the top ten scores overall
Top score on each band for single band entries
Top score in each state, province, and DXCC country.
Information and logs:
Red Reynolds, K5VOL
835 Surryse Road
Lake Zurich, IL 60047

MICHIGAN QRP CLUB CW CONTEST
Sponsor
Michigan QRP Club.
Dates
Third weekend in January.
Contest Period
1200 UTC Saturday to 2400 UTC Sunday.
Objective
Everybody works everybody.
Bands
160-6 meters.
Type of Competition (QRP)
CW Only: Class A (less than 250 milliwatts)
Class B (less than 1 watt)
Class C (less than 5 watts).
Exchange
RST, QTH, MI QRP number or power output for nonmembers.
Points
Five points for member QSO and one point for nonmember QSO.
Multipliers
States, provinces, DXCC countries per band.
Scoring
(Total QSO points) × (sum of multipliers per band) × (1.25 for battery power or 1.5 for natural power).
Awards
Certificates
Information and Logs
L.T. Switzer, N8CQA
654 Georgia
Marysville, MI 48040

UBA CONTEST

Sponsor
Union of Belgian Amateurs
Dates
CW: Fourth weekend in January
SSB: Fourth weekend in February
Contest Period
1300 UTC Saturday to 1300 UTC Sunday.
Objective
Everybody works Europe.
Bands
80-10 meters.
Type of Competition (QRP)
Single operator, all band.
Exchange
RS(T) and consecutive serial number. Belgium stations also send province abbreviation.
Points
Ten points for ON, DA1, DA2 QSO
Three points for other European community members
One point for any other QSO
Multipliers
Belgian provinces: AN, BT, HT, LB, LG, LU, NR, OV, WV
Each of the prefixes: ON4-ON9, DA1, DA2
European communities: CT, CU, DL, EA, EA6, EI, F, G, GD, GI, GJ, GM, GU, GW, I, IS, LX, OZ, OY, PA, SV, SV5, SV9, SY, TK, ZB2
Scoring
Total QSO points times total multipliers
Awards
Certificates
Information and Logs
UBA HF Contest Committee
Glaicia Jan, ON6JG
Oude Gendarmeriestraat 62
B-3100 Heist Op Den Berg
Belgium

ARRL MEMBERS
This proof of purchase may be used as a $0.60 credit on your next ARRL purchase. Limit one coupon per new membership, renewal or publication ordered from ARRL Headquarters. No other coupon may be used with this coupon. Validate by entering your membership number—the first 7 digits on your QST label—below:

YOUR QRP OPERATING COMPANION

PROOF OF PURCHASE

FEEDBACK

Please use this form to give us your comments on this book and what you'd like to see in future editions.

License class:
☐ Novice ☐ Technician ☐ Technician with HF privileges
☐ General ☐ Advanced ☐ Extra

Name _____	ARRL member? ☐ Yes ☐ No
_____	Call sign _____
Daytime Phone ( ) _____	Age _____

Address _____
City, State/Province, ZIP/Postal Code _____
If licensed, how long? _____
Other hobbies _____

For ARRL use only
Edition 1 2 3 4 5 6 7 8 9 10 11 12
Printing 1 2 3 4 5 6 7 8 9 10 11 12

Occupation _____

From _____
_____
_____

Please affix postage. Post Office will not deliver without postage.

EDITOR, YOUR QRP OPERATING COMPANION
AMERICAN RADIO RELAY LEAGUE
225 MAIN ST
NEWINGTON CT 06111

please fold and tape